ENCLOSURE MASONRY WALL SYSTEMS WORLDWIDE

 BALKEMA – Proceedings and Monographs in
Engineering, Water and Earth Sciences

Enclosure Masonry Wall Systems Worldwide

Typical Masonry Wall Enclosures in Belgium, Brazil, China, France, Germany, Greece, India, Italy, Nordic Countries, Poland, Portugal, The Netherlands and U.S.A.

Organized by

CIB W023 – WALL STRUCTURES

CIB Publication

Editor
S. Pompeu Santos

Taylor & Francis
Taylor & Francis Group

LONDON / LEIDEN / NEW YORK / PHILADELPHIA / SINGAPORE

Taylor & Francis is an imprint of the Taylor & Francis Group, an informa business

© 2007 Taylor & Francis Group plc, London, UK

Typeset by Charon Tec Ltd (A Macmillan Company), Chennai, India
Printed and bound in Great Britain by Bath Press Ltd (CPI-Group), Bath

Published by: Taylor & Francis/Balkema
            P.O. Box 447, 2300 AK Leiden, The Netherlands
            e-mail: Pub.NL@tandf.co.uk
            www.balkema.nl, www.taylorandfrancis.co.uk, www.crcpress.com

British Library Cataloguing in Publication Data
A catalogue record for this book is available from the British Library

Library of Congress Cataloging in Publication Data
Enclosure masonry wall systems worldwide : typical masonry wall enclosures in Belgium and the Netherlands, Brazil, China, France, Greece, India, Italy, Nordic Countries, Poland, Portugal, and U.S.A. / organized by CIB W023–Wall Structures ; edited by, coordinator, S. Pompeu Santos.
       p. cm.

   Includes bibliographical references and index.
   ISBN-13: 978-0-415-42577-3 (hardcover : alk. paper)
   ISBN-10: 0-415-42577-8 (hardcover : alk. paper)
  1. Exterior walls–Design and construction–Case studies. 2. Brick walls–Design and construction–Case studies. 3. Building–Cross-cultural studies. I. Santos, S. Pompeu.

TH2235.E52 2006
693′.1–dc22

                                    2006028600

ISBN13: 978-0-415-42577-3 (hbk)

Table of Contents

Preface

S. POMPEU SANTOS

Research Officer
LNEC
Lisbon
Portugal

CIB W023 Coordinator

Enclosure walls have a key role in building construction, providing structural safety, protection against intrusions and comfort, among other features.

Enclosure walls solutions are traditional, by nature, but they follow the development and progress in the construction activity sector in each region or country. Hence, there is a large variety of solutions around the world for enclosure walls in buildings, depending on the local conditions, the climate and the building technologies more easily available in the region.

Enclosure walls had a great move in recent times influenced by the development of new standards at national and international levels. In this issue, the new European Standards related to masonry (Eurocode 6 and product standards) should be mentioned. In some countries, new demands, namely on thermal insulation or earthquake resistance, are also leading to the development of new materials and wall solutions.

At a moment when globalisation becomes increasingly widespread, the importance of knowing the present situation about the enclosure masonry walls around the world becomes evident, namely, the used solutions, the existing problems and the trends on evolution.

Therefore, the Commission W023 – Wall Structures of CIB – International Council for Research and Innovation in Building Construction, which deals with walls in their different functions, decided to prepare a publication presenting the situation of the enclosure masonry walls used in a selected group of countries or regions that could give a good picture of the situation in the different parts of the world.

This volume is constituted by several contributions from W023 Commission members and invited experts with recognised curricula and knowledge about the situation in their countries. The following countries/regions are represented: Belgium, France, Germany, Poland and The Netherlands, from Central Europe; Greece, Italy and Portugal, from South Europe; Nordic Countries, from North Europe; Brazil and U. S. A., from the Americas; and, China and India, from Asia.

Contributions to this volume have a similar format, covering the following items:

- characterisation of the building sector in each country/region (importance of the building sector, structural systems used in buildings, etc.);
- overview of the most common masonry materials (units, mortar, reinforcement, thermal insulation, wall finishes, etc.);
- description of the typical enclosure wall solutions and the existing problems; and
- trends on evolution (resolution for existing problems and new developments).

I would like to thank the authors, both the Commission members and those from outside of the Commission, for their contributions for the book and for believing in its importance and feasibility. I also thank Josep M. Adell, who had the initial idea of preparing such a publication. I would also like to thank CIB and Balkema for publishing the book.

I hope that the book will be fruitful and contribute for the development and the progress of the enclosure masonry wall systems around the world.

Lisbon, May 2006

CHAPTER 1

Typical masonry wall enclosures in Belgium and The Netherlands

Dirk R.W. MARTENS

Professor Ir.-Arch.
Eindhoven University of Technology
The Netherlands

SUMMARY

In Belgium and The Netherlands the majority of the enclosure walls for housing buildings are designed as cavity walls with clay brick masonry veneers. Also concrete bricks and blocks and calcium silicate bricks are sometimes used for veneers. In Belgium the inner wall leaves are mostly built in clay brick masonry and in a lesser extent in concrete, AAC or calcium silicate. In The Netherlands the majority of the load bearing inner walls consists of calcium silicate elements with thin layer mortar.

Generally masonry walls are built on site which is very labour intensive. Due to the lack of skilled masons during the last two decades research has been focused on prefabrication and automation for the production of masonry walls. Thin layer mortar masonry and dry stacked bricks and blocks are new building techniques that become more and more popular.

1 INTRODUCTION

Traditionally Belgium and The Netherlands are countries where most buildings are made of masonry, in particular for housing. In Belgium nearly every one-family house is made of clay bricks and blocks while in The Netherlands calcium silicate elements are used more frequently for the load-bearing inner walls combined with clay brick veneers.

In this report a concise comparison between the building traditions in these two countries will be presented. Moreover a description of the existing problems with masonry enclosure walls will be given and finally some new developments are dealt with.

2   BUILDING SECTOR

2.1   *Importance of building sector*

In Belgium and The Netherlands the building sector is quite important in particular in the field of employment. At this moment only limited statistical information is available but, in general, it can be stated that the building industry accounts for 5 to 10% of the total economical activity (Figure 1).

In Figure 2 an overview is given of the yearly production of clay bricks and blocks, which is an indication of the importance of the clay brick industry in both countries. In Belgium almost 60% of the production consists of perforated blocks for load-bearing inner walls. In The Netherlands only clay bricks for veneers are produced. In 2002 this production accounted for, approximately, 250 million euro.

2.2   *Structural systems for buildings*

Structural systems for buildings in Belgium and The Netherlands consist mainly of reinforced concrete frames and walls or load-bearing masonry or a combination of both. Timber structures are not used frequently and steel frames are used for

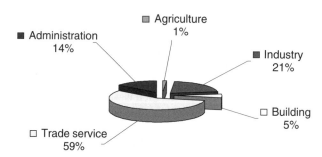

Figure 1.   Importance of building industry in Belgium in 2001.

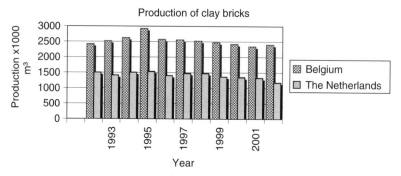

Figure 2.   Production of clay bricks in Belgium and The Netherlands [1]–[2].

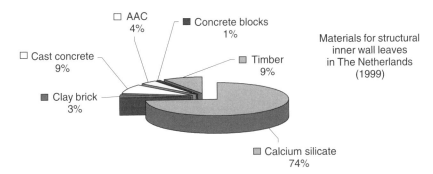

Figure 3. Importance of the different masonry materials for structural systems in buildings in The Netherlands.

special building typologies as industrial buildings, railway stations, high-rise office buildings, etc.

In The Netherlands the majority of the load-bearing walls are built in calcium silicate elements or blocks (Figure 3). This is not the case in Belgium where most of the walls are built in perforated clay blocks and in a lesser extent in concrete blocks and where the share of calcium silicate elements and blocks is less important.

2.3 *Masonry enclosure systems used for housing*

In wintertime it is quite cold in Belgium and The Netherlands while in summertime it often rains. These climatic circumstances have led to the introduction of the cavity wall since the beginning of the 20th century. In this system, which became widespread since the 1960's, the watertightness is secured by the outer wall leaf.

For sustainability reasons, requirements concerning thermal insulation of buildings have become more and more severe which has resulted in an increasing thickness of the thermal insulation introduced in the cavity walls.

For single leaf walls the watertightness is realised by rendering or cladding and the thermal requirements are met by applying thermal insulation or by using lightweight masonry blocks. As a result of this evolution the following enclosure systems for housing buildings are encountered in Belgium and The Netherlands:

(1) cavity wall with air cavity not or partly filled with thermal insulation (air cavity of at least 30 mm) and both wall leaves in masonry (Figure 4);
(2) cavity wall with cavity completely filled with thermal insulation and both wall leaves in masonry;
(3) single leaf masonry wall with rendering: in this case external thermal insulation or light weight bricks and blocks are used to meet the thermal insulation requirements;
(4) single leaf masonry wall with cladding (tiles, plastic, wooden or metallic cladding, etc …) combined with external thermal insulation or light weight bricks and blocks;
(5) cavity wall with cavity partly filled with thermal insulation and with concrete or timber structural inner wall leave and masonry veneer.

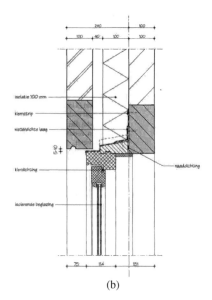

(a)   (b)

Figure 4.   (a) Cavity wall in Belgium with cavity of 90 mm partly filled with 40–60 mm of thermal insulation: inner wall leaf in perforated light weight clay blocks (140–190 mm) and clay brick veneer (90–100 mm) [3] (b) Cavity wall in The Netherlands with cavity of 90–140 mm partly filled with 60–100 mm of thermal insulation: inner wall leaf in calcium silicate elements (100–120 mm) and clay brick veneer (100 mm).

In The Netherlands and the Flemish part of Belgium system (1) is used most frequently while in the southern part of Belgium also system (3) is popular. All the described systems are executed using the traditional building technique on site. In The Netherlands only prefabricated mortar is used. In Belgium masonry walls are sometimes built with site-made mortar, in particular for smaller buildings.

3   MOST COMMON ENCLOSURE WALL SYSTEMS

3.1   *Masonry materials*

a)   Masonry units
Besides the fact that, in Belgium and The Netherlands most enclosure walls are built in masonry, there is a remarkable difference in the used building materials. For inner (load-bearing) wall leaves in Belgium the following materials are used (in descending order of importance):

– perforated clay blocks (Figure 5);
– concrete blocks (normal weight and light weight) (Figure 6);
– autoclaved aerated concrete blocks (Figure 7);
– calcium silicate blocks (Figure 8);
– calcium silicate elements (Figure 9).

Figure 5.   Different types and sizes of perforated clay blocks used in Belgium [3].

Figure 6.   Solid and hollow concrete blocks for load-bearing masonry.

In The Netherlands, nearly no clay bricks and concrete bricks or blocks are used for structural masonry but the majority of the inner wall leaves are built in calcium silicate elements (Figure 9). A minor part is realised with calcium silicate blocks. Since the shrinkage of calcium silicate is quite large, movement joints are required with a spacing of approximately 6 to 8 m.

For veneer walls in both countries almost the same materials are applied as can be seen in Table 1. In The Netherlands more than 90% of all veneer walls are made of clay brick masonry (Figure 10). In Belgium the share of concrete brick and block masonry is a somewhat more important but still limited (Figure 11).

Figure 7.   Autoclaved aerated concrete blocks.

Figure 8.   Calcium silicate blocks.

Figure 9.　Inner wall leaves in calcium silicate elements in The Netherlands.

Table 1.　Materials used for veneer walls in Belgium and The Netherlands in descending order of importance.

Belgium	The Netherlands
Clay bricks	Clay bricks
Concrete bricks and blocks	Concrete bricks
Calcium silicate bricks	Calcium silicate bricks
Natural stone blocks (southern part of Belgium)	

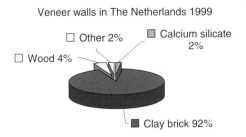

Figure 10.　Materials for veneer walls in The Netherlands.

b)　Mortars

In Belgium, for small buildings, masonry mortars are still made on site. For more important building sites prefabricated mortars are more common. In most cases these factory-made mortars are dry mortars and are delivered on site in silos.

Usually a mix of Portland cement (CEM I 32.5) and sand is used but in some cases a lime-cement-mortar is prescribed. Admixtures to improve the workability or to improve the water-repellent behaviour are used frequently.

The situation in The Netherlands is quite comparable. The only difference concerns the use of more cement mortars instead of lime-cement mortars. Moreover, in The Netherlands almost no site-made mortars are used anymore.

Figure 11.   Veneer wall in thin layer mortar masonry with concrete bricks in Belgium.

Figure 12.   Typical wall ties used in Belgium and The Netherlands.

c) Ancillary components

The use of wall ties in galvanised or stainless steel is common practice in Belgium and The Netherlands.

These ties provide stability to the slender wall leaves during building and transmit wind loading from the outer wall leaf to the inner wall leaf. Depending on the height of the building and the exposure to wind loading, 4 to 6 wall ties per m^2 with a diameter of 4 or 5 mm are required (Figure 12).

Lintels over windows and doorways are usually made of steel or concrete. In some cases reinforced or prestressed masonry lintels are used.

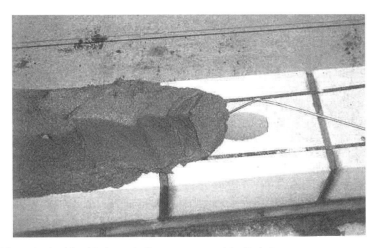

Figure 13.   Typical bed joint reinforcement used in Belgium.

The spacing of movement joints in veneer walls is different in Belgium and The Netherlands. For clay brick veneers in Belgium a maximum distance of 30 m between the movement joints is allowed while in The Netherlands the spacing is limited to 12 m.

Prefabricated bed joint reinforcement of the truss type in galvanised or stainless steel (Figure 13) is used frequently in Belgium but not in The Netherlands.

Reinforcing the masonry is meant to enhance the moment capacity of a masonry beam or wall, or to limit the crack width due to differential settlements or thermal movements.

In Belgium the spacing of the movement joints in veneer walls may be enlarged by 50% if bed joint reinforcement is applied. This is not allowed in The Netherlands.

3.2   *Thermal insulation*

Since the U-value (thermal transmission factor) of enclosure walls in Belgium and The Netherlands is limited to respectively 0.60 W/m^{2}K, and 0.40 W/m^{2}K, it is nearly impossible to build enclosure walls without thermal insulation.

In cavity walls the space between the two leaves are filled partly or completely with thermal insulation. For single leaf walls, external insulation is required. Only when very thick wall leaves are used combined with units with improved thermal performance, it is possible to build enclosure walls without thermal insulation.

The following thermal insulation materials are applied in both countries:

– mineral wool (semi-hardboard) (Figure 14);
– polyurethane foam;
– extruded polystyrenes (Figure 14);
– various biological insulation materials.

3.3   *Damp proof courses*

Damp proof courses are required in every masonry building in Belgium.

(a)   (b)

Figure 14.   Thermal insulation with: (a) mineral wool and (b) extruded polystyrenes.

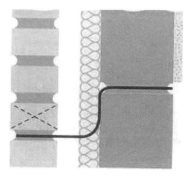

Figure 15.   Damp proof course in cavity walls in Belgium [3].

Different varieties of polyethylene DPC's are introduced during the last decades. In some cases rubber, asphaltic sheets or lead are still used. These damp proof courses are not only used in both leaves of the outer walls but are also introduced in the inner walls (Figure 15).

In The Netherlands the water suction is prevented by a rich hydrofuge mortar in the lower courses and almost no DPC-sheets are used.

3.4   *Wall finishes*

As stated earlier, for housing buildings, the majority of the enclosure walls are finished with a clay brick veneer. In the other cases these walls are finished with renderings, ceramic tiles, natural stone tiles, wooden sidings or metal sheeting.

3.5   *Most common masonry buildings enclosures*

The most common masonry enclosure walls in Belgium and The Netherlands are presented in Figure 4.

3.6 *Existing problems*

a) Architectural design

The architectural design of the enclosure walls, including the choice of materials, is always done by the architect based on his personal experience and preferences. If structural problems are encountered a structural engineer will be hired.

    Concerning masonry veneer walls, most problems arise due to incorrect detailing of the masonry:

- staining on the wall due to preferential drainage of rainwater;
- too small spacing of movement joints which are not integrated in the architectural design;
- efflorescences and moss growth caused by incorrect detailing of sills and wall copings;
- incorrect positioning of damp proof courses;
- restrained differential deformation and cracking of veneer walls due to rigid connection of inner wall leaves to outer wall leaves.

b) Structural design

Problems in the field of structural design are caused by a lack of knowledge:

- oversized or undersized wall thicknesses by lack of proper calculation;
- cracking of masonry walls due to shrinkage, settlements of foundation or deformation of supporting concrete slabs and beams or metal girders;
- incorrect design of lintels (self-supporting or integrated lintels).

c) Execution

Lots of problems are not only caused by the architectural or structural design but also by incorrect execution of the masonry, sometimes due to bad communication between designer and contractor. This problem is continuously enlarged by the still growing lack of skilled masons.

4   EVOLUTION

4.1 *Solution of existing problems*

The solution of the problems concerning architectural and structural design should be provided by education of architects and engineers and by dissemination of the requirements stated in the codes and other regulations.

    Solving the problem of the lack of skilled masons requires the development of new building techniques.

4.2 *Trends regarded to labour, quality, durability and productivity*

The improvement of labour working conditions combined with growing productivity and the improvement of the quality of masonry can only be realised by the introduction of prefabrication, automation and robotics. Long time there has been no effort in this field, but nowadays the masonry industry in Belgium and The Netherlands has realised that research should be focused on these topics in order to secure the future of masonry.

Figure 16.   Thin layer mortar masonry.

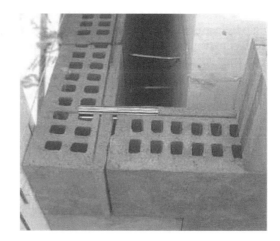

Figure 17.   Dry stacking technique (click brick).

4.3   *New developments*

As a result of the research that has been carried out during the last two decades particularly in The Netherlands, some new building techniques have been developed which are now introduced in the building market:

Figure 18.   Prefabricated sandwich-panels of masonry and concrete.

Figure 19.   Prefabricated veneer walls of clay brick (noise barrier).

(1) thin layer mortar masonry for veneer walls: use of a mortar spaying gun or mortar tray and putting the bricks without a trowel (Figure 16);
(2) dry stacking technique for veneer walls: no mortar is put between the bricks but special clips in stainless steel provide the mechanical connection between the different bricks (Figure 17);
(3) prefabrication of cavity walls: combination of inner wall leaf in reinforced concrete with outer wall in masonry and insulation in between (sandwich-panels) (Figure 18);
(4) prefabrication of veneer walls: veneer walls in thin layer mortar masonry with safety reinforcement (Figure 19);

Figure 20.   Perforated clay block masonry with thin layer mortar and unfilled tongue and groove perpend joints.

(5) thin layer mortar clay block masonry for load-bearing inner walls with unfilled perpend tongue and groove joints (Figure 20).

These new techniques are being developed further at this moment and it is hoped that they will replace the traditional masonry methods step by step thanks to the reduction of the cost price and the improvement of the quality of the masonry.

5   REFERENCES

[1] KNB – "Jaarverslag KNB 2002". De Steeg, maat 2003 (in Dutch).
[2] BBF – "Baksteenproductie zit opnieuw in de lift". Bouwkroniek, 1 Augustus 2003 (in Dutch).
[3] VZW Bouwen met baksteen – "Handboek snelbouwbaksteen". Brussel, 1997 (in Dutch).
[4] Martens, D. R. W., Vermeltfoort, A. T., Bertram, G. – "Ontwerpen en dimensioneren van Metselwerkconstructies". Dictaat TU Eindhoven, uitgave 2002 (in Dutch).
[5] Nationaal Instituut voor de Statistiek (NIS) – "website Augustus 2004" (in Dutch).

CHAPTER 2

Typical masonry wall enclosures in Brazil

Humberto
R. ROMAN

Professor UFSC
Florianópolis SC
Brazil

Denise
A. da SILVA

Professor UFSC
Florianópolis SC
Brazil

SUMMARY

Masonry is the most used enclosure wall systems in Brazilian buildings, either structural or non-load bearing masonry. The walls are made of ceramic, concrete or calcium silicate bricks and blocks. The present paper describes the materials, techniques of construction and durability aspects of Brazilian masonry walls. Some aspects of the Brazilian construction industry are also pointed out.

1   INTRODUCTION

The Construction Industry is very much important for Brazilian economy, responding for about 6% of the Gross National Product. Despite this importance, the quality of the construction standards is not necessarily good. Construction waste is said to be about 20%. Generally, the problems are attributed to bad workmanship. However, some research works have pointed errors along the whole building process. The main causes of errors in Brazilian construction process are shown in Figure 1.

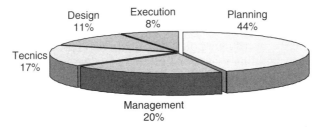

Figure 1.   Origin of errors in Brazilian construction industry.

During the seventy and eighty decades of last century, the companies generally constructed for upper and middle class which used government mortgages to buy. Huge inflation and recession cause the dismantlement of the system and companies had to produce cheaper buildings to keep themselves in the market. On the otherhand, the deficit of 6.5 million houses for lower classes makes this market very attractive. The challenge was to produce flats and houses affordable for these consumers.

In 1997, the Brazilian government launched the Brazilian Quality & Productivity Program for Habitat – PBPQ-H. PBQP-H has been a voluntary quality program for the construction sector that has had the government and the private sector working together for quality improving and adopting a vision of Productive Network. The major objective is to promote improvement of the quality and productivity in the construction sector to increase produced goods and services competitively.

The specific objectives of the Program are improve quality of management at construction companies and in the projects; to improve quality of materials, components and construction systems; to enhance structural development and disseminate technical standards, practical and construction codes; to introduce instruments and mechanism for improving quality of projects; to stimulate the creation of programs to set and training labour and to support introduction of technological innovations.

At the present time, there are more than 1548 construction companies officially taken part of the program, which is leading to increasing quality of buildings.

This report presents the situation of Brazilian construction regarding to the wall enclosures.

2   STRUCTURAL MASONRY

2.1   *Structural masonry in Brazil*

The use of structural masonry has growing very fast in Brazil during the last decades, despite the few university courses that offer lectures on this subject.

The first structural masonry buildings were built in the late sixties. They were made in concrete blocks and had a very poor architectural appeal, as can be seen in (Figure 2a). The first high rise building built in S. Paulo was on reinforced blockwork. It is 12 storeys high and was designed for a North-American engineer (Figure 2b).

For many years, the procedures to build on structural masonry were very similar to the ones used for conventional buildings. The potential advantages of structural masonry were not achieved. At the beginning of the nineties, the decrease of inflation along with government policies of quality and productivity increased the use of this type of building construction process. Changes have been noticed on quality of components, equipments, tools, materials and workmanship (Figure 3).

After 1990 the use of un-reinforced concrete structural blockwork became very common in the Southeast and South of Brazil. Concrete blocks with 15 cm module were more often used. Some structural clay blocks also started to be produced. The use of pre-moulded stairs, appropriated equipments and site control

(a)   (b)

Figure 2.   (a) Un-reinforced structural masonry from late sixties; (b) First high-rise building on structural masonry in S. Paulo.

(a)   (b)

Figure 3.   Changes on the construction process.

increased the rationalization of the process. Some polemic innovations appeared, specially the use of un-filled vertical joints on very tall buildings. Figures 4 and 5 show common types of unreinforced structural masonry buildings, currently built in Brazil.

Figure 4.   Unreinforced masonry buildings.

Figure 5.   Twelve and fourteen storey high un-reinforced blockwork buildings.

2.2   *Materials*

a)   Masonry units

The masonry units most used are concrete and clay blocks. Very few companies use solid clay bricks or silicate-lime blocks.

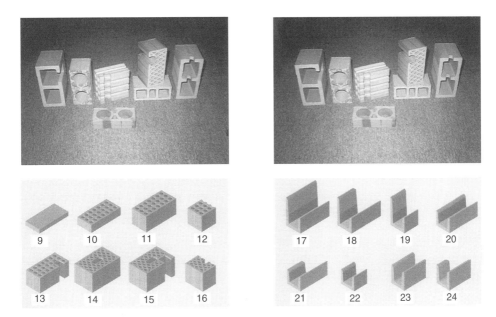

Figure 6.   Some clay blocks produced in Brazil.

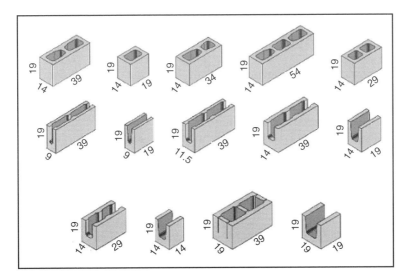

Figure 7.   Different types of concrete blocks.

Special types of blocks are used to build lintels and the last masonry course. Also blocks designed for electric and hydraulic connections are used. Figures 6 and 7 show the main types of blocks found in the market.

The compressive strength of the clay blocks produced is normally around $10 \, N/mm^2$. However in the South of Brazil it is possible to find clay blocks with strength up to $25 \, N/mm^2$ on gross area.

Different types of concrete blocks are produced, normally with 15 cm or 20 cm modules. These can be seen in Figure 7. Brazilian standard for concrete blocks divide the blocks according to the compressive strength in different classes, ranging from 4.5 to 16 N/mm^2 on gross area.

The quality of both types of masonry units, ceramic and concrete, varies very much. It is possible to find products of very high quality and others of very poor production control.

b)   Mortar and grout

Joints are filled normally with cement:lime:sand mortars. Portland cements are used in Brazil for the mortars production. Besides clinker and gypsum, they may contain calcium carbonate, fly ash or ground blast furnace slag as admixtures. The sand is usually from natural deposits, as watercourses and dunes, but the use of artificial sand (from rocks crushing) is common.

Many builders add hydrated lime to the mortars to improve mortars properties as workability, water retention, drying shrinkage and strain. The use of air entraining agents instead of hydrated lime is also common. These are wrongly considered as "hydrated lime substitutes" in Brazil.

The materials proportions commonly used for structural masonry mortars production range between 1:1/2:4.5 and 1:1 or 2:6 (cement:hydrated lime:sand, in volume). When air-entraining agents are used, they are pre-mixed to the water in the quantity recommended by manufacturer.

2.3   *Structural design of masonry*

The main difference between concrete framed and structural masonry designs is that the later must be much better detailed. Whilst the design of concrete framed buildings normally shows concern only with structural matters, leaving most of the decisions to the site, in general, structural masonry design show detailed solutions for hydraulic, electrical and other services.

The "executive" set of design plants includes: location of the walls, first and second course plan, pagination of every wall, position of steel inside the blocks, details of lintels and last course reinforcement, location of precast elements and so on.

The best designs show also which are the walls possible to be removed in case of a later refurbishment. Figures 8 and 9 show examples of design detailings of buildings.

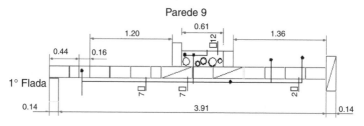

Figure 8.   Shaft for hydraulic elements.

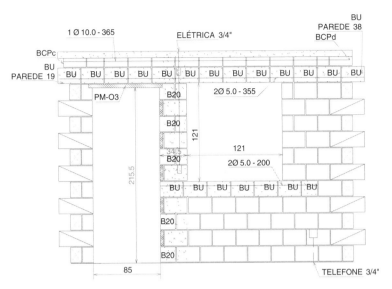

Figure 9.   Example of pagination.

2.4   *Construction of structural masonry buildings*

Many companies that used to build concrete reinforced structures, when moving to structural masonry adopt a more rationalized procedure of construction. Besides, a more detailed design, special equipments and workmanship training programs are introduced to the building process. The equipments are quite simple but improve very much the control of joints thickness, plumb and construction speed. Some of them are shown in Figure 10.

The main procedures to build the walls are:

(1) Positioning of first course strategic blocks on a levelled slab starting at the slab highest level (Figure 11a and b).
(2) Completion of first course (Figure 11c);
(3) Positioning of metallic corner (Figure 11d);
(4) Laying of the following courses.

Figure 12 shows a general view of a rationalized structural masonry site.

3   NON LOADBEARING MASONRY WALLS

Basically, the materials used for the construction of non-structural masonries in Brazil are unit bricks or blocks[1] with cement-based mortars.

a)   Masonry units
There is a large variety of masonry units available in the Brazilian market. Natural rocks, ceramic bricks and blocks, concrete blocks, calcium-silicate blocks, aerated

---

[1] Bricks are bulky units, while blocks are units with holes.

Figure 10. (a) Metallic corner used to keep verticality and level; (b) metallic mortar container; (c) "German" level; (d) plumb and level rule; (e) "fork" carrier; (f) tube.

concrete bricks/blocks, glass bricks and cement-soil bricks are currently used for the construction of non-structural masonries.

The choice is made mainly based on local availability of the products. On special cases the bricks/blocks characteristics, such as the weight, the shape and dimensions, the position and geometry of the holes, the surface texture and the

(a)

(b)

(c)

(d)

Figure 11.   Construction of structural masonry buildings.

Figure 12.   Aerial view of a structural blockwork site.

physical, chemical and mechanical characteristics and properties of the bricks/blocks are also considered.

All of these characteristics and properties must be in accordance with the conditions to which the masonry will be exposed during its service life. By far the most used units are ceramic and concrete blocks, for both external and internal walls. Exception is made for fire resistant walls, which are mostly built with aerated concrete bricks/blocks due to their high fire propagation resistance.

Ceramic blocks, which are used in more than 90 percent of the Brazilian buildings, are extruded, and can have 2, 4, 6, 8 or 21 holes. The blocks are layered with the holes in the horizontal direction, except the 21 holes ones. The compressive strength must be at least $1.0\,N/mm^2$, according to the Brazilian Society for Technical Standards (ABNT).

Concrete blocks are also used but in much less proportion. The ABNT codes establish the minimum compressive strength of $2.5\,N/mm^2$ for non-structural concrete block.

Autoclaved-aerated concrete bricks/blocks are mostly used for fire-barrier walls, generally at the stair region forming the emergency exits of the buildings. When better acoustic and heat insulation is needed, aerated concrete bricks are used for the construction of external walls. Besides cement, lime and sand in their composition, the bricks/blocks are produced with aluminium powder, which reacts in the mixture and releases a large amount of small gas bubbles into the concrete. For the cure, the bricks/blocks are heated in an autoclave under the pressure of 10 atm at 180°C.

In spite of the recommendations of Brazilian standards for ceramic and concrete bricks/blocks, generally there is no site control for these materials acceptance, resulting in a huge waste of materials and human labour. However, the governmental programme for quality and productivity on the construction industry, PBQP-H, has brought improvements on services and materials control. The later may be done by visual analysis and/or laboratory tests and it is a criterion to accept or reject most of the building material delivered at the construction site.

b)   Mortar

As for structural masonry cement and sand mortars are generally used for non load-bearing walls. The material proportion is not very well controlled, but normally it ranges from 1:1:6 to 1:2:9 (cement:lime:sand).

Some builders use to replace hydrated lime for clay soil to the joint laying mortars, since clay is cheaper and it also improves mortars plasticity. However, experts do not recommend this procedure because many durability problems may happen at the building due to the presence of unburned clay in the mortars.

c)   Joints and reinforcement

The structure of most Brazilian buildings is of concrete reinforced frame composed by beams, columns and slabs. Also the foundations are built in reinforced concrete. In most of these buildings, non-structural masonries are used as enclosure wall (Figure 13).

Although not being structural, the walls showed in Figure 13 may be submitted to significant efforts due to the connection with the reinforced concrete frame. Short and long-term concrete frame deformations (drying shrinkage, elastic and creep deformations and movements caused by the wind); foundations and thermal movements are some of the causes of masonry stresses. These stresses are the major cause of micro cracks development in the walls.

Furthermore, drying shrinkage of the mortar produces volumetric decrease of the masonry. Depending on the length and height of masonry, and also on the

Figure 13.   Example of reinforced concrete frame and ceramic block masonry enclosure walls in Brazilian building.

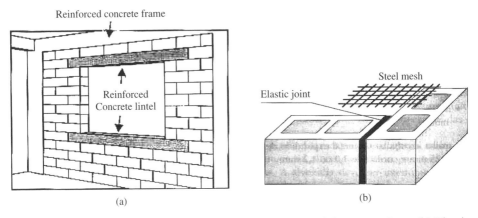

Figure 14.   (a) Reinforced concrete lintels at windows and doors openings; (b) Elastic joint connecting two masonry walls.

mortar characteristics, higher masonry deformation occurs, and hence, cracks. In spite of those, there is an apparently irreversible tendency to design more flexible concrete frames, which increases both the short and long-term deformations.

The new challenge of Brazilian designers and builders has been to adequate concrete frame and masonries deformations to minimize the occurrence of cracks and distortions. Some solutions have been used, such as movement joints and reinforced concrete lintels. In addition, the connection between the masonry and the concrete frame has been adjusted to the expected deformations. All these artefacts (examples are shown in Figure 14) have improved the performance and durability of the Brazilian buildings.

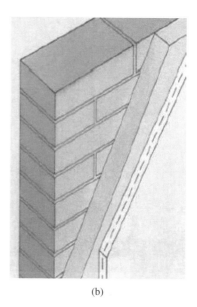

(a)   (b)

Figure 15.   (a) Rigid glass wool panels used as insulating material in a double masonry wall; (b) Illustration of polystyrene layer beneath finishing.

4   THERMAL AND ACOUSTIC INSULATION OF MASONRY WALLS

The thickness of the external walls is normally between 100 mm to140 mm all over the country with exception of the Rio Grande do Sul State where the minimum external wall thickness is 250 mm in many cities.

Brazilian territory extends from just above the Equator up to latitude 15° in the South Hemisphere. The climate varies from sub-tropical to equatorial, and it is predominantly hot. Brazilian standards and codes do not specify values for thermal or acoustic insulation of the walls.

However, during the last few years, due to the lack of energetic power suppliers, Brazilian Government has made many attempts to reduce the consumption of energy, including some economic penalties to the consumers who exceed an established quantity of kVA/month.

The energetic sector crisis has lead designers and constructors to improve the internal thermal comfort. Many insulating materials are now available in the Brazilian market, and many designers and constructors are concerned with the need of energy savings. Rigid glass wool panels and polystyrene layers are the insulating materials most used in Brazilian masonry walls.

Figure 15 shows both systems. However, hot and acoustic insulated walls still are not very much in use.

5   DAMP-PROOFING MATERIAL

Brazilian standards require that a damp-proofing layer or sheet must cover any reinforced concrete beams at the ground level to avoid the rising of water by capillary forces. Furthermore, the mortar used to lay the masonry units must have a hydrofuge additive. In addition, the external wall face rendering must have impermeable additive in the first 500 mm from the ground level in order to avoid the infiltration of spilling rainwater into the wall.

6 WALL FINISHES

Generally, the masonry walls of Brazilian buildings are finished with cement-based mortars.

The most common mortars are the following: (i) Portland cement, hydrated lime and sand, produced in situ; (ii) Portland cement, sand and air-entraining agent, also produced *in loco*; (iii) pre-packed dry-set mortars, mixed with water at the building site; (iv) ready-to-use mortars, delivered by mixer-trucks at the building site.

The thickness of the rendering is around 25 mm externally and 20 mm on internal walls. This thickness may become significantly higher (up to 70 mm) to correct the wall plumb or may decrease to around 5 mm on internal walls of rationalized buildings process, especially in structural concrete blockwork.

In some areas of Brazil the use of gypsum plaster for internal wall rendering is very common. They are gypsum-sand based mortars, and, sometimes, hydrated-lime and chemical additives are added to the mixture to improve its workability and to control the setting time. As the cement-based mortars, gypsum mortars are manually (Figure 16) or jet-applied (Figure 17).

The external finish of the render is usually an organic paint layer or a ceramic wall tile system. The later is mainly used to cover bathrooms, kitchens and laundries walls and floors.

However, the use of ceramic tiles in other building walls (Figure 18a) and in façades has been increasing all over the country, since Brazil is one of the biggest producers of ceramic tiles in the world.

Different finishing systems are sometimes applied to masonry walls, such as: (i) rendering made with medium and fine aggregates embedded in plastic resin (Figure 18b); (ii) aluminium panels connected to the wall by metallic inserts; (iii) rock panels or ceramic tiles also connected to the wall by metallic inserts.

(a)

(b)

Figure 16.   Manually applied gypsum mortar.

Figure 17.   Jet-applied gypsum mortar.

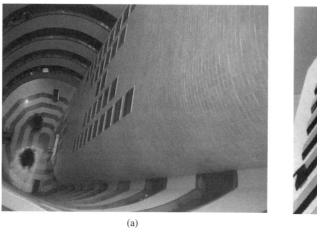

(a)

(b)

Figure 18.   (a) Large masonry wall covered by ceramic tiles in internal environment; (b) Aggregate-resin rendering.

7   NEW DEVELOPMENTS

The tradition in Brazilian construction is to use clay blocks on non-loadbearing walls. The general bad quality of these blocks is, perhaps, the main cause of bad quality of non-loadbearing walls.

The walls are very often design in a non-modular way due to the difficulty to find modular blocks in the market. Higher amounts of mortar have to be used to level the wall courses due to both blocks dimensional variations and high block suction rate. Lack of plumb is responsible for thicker mortar rendering, increasing the surface cracks.

As mentioned before, the governmental program for quality and productivity, PBQP-H, has been helping to change the construction scenery. Moreover, the appearance in the market of gypsum-based dry wall has made the brick and block industry look for quality products improvements and development of new types of products. Dry walls are still not very well accepted by the market but its use is continuously growing.

Some companies have developed building systems based on clay blockwork panels. These systems have still been used for one for one-storey houses. Research program for multi-storey structural masonry panel is being run.

8   REFERENCES AND DOCUMENTATION

[1] ABCI – "Associação Brasileira de Construção Industrializada". Manual Técnico de Alvenaria, Projeto, Editores Associados Lda, Editor Vicente Wissenbach.1990, 275 p. (in Portuguese).

[2] MENDES, R. K. – "Resistência à Compressão de Alvenaria de Blocos Cerâmicos Estruturais". Dissertação de mestrado, Curso de Pós-Graduação em Engenharia Civil, UFSC, Agosto de 1998 (in Portuguese).

[3] MOHAMAD, G. – "Comportamento Mecânico na Ruptura de Prismas de Blocos de Concreto2. Dissertação de mestrado, Curso de Pós-Graduação em Engenharia Civil, UFSC, Maio de 1998 (in Portuguese).

[4] NBR 12118 – "Blocos vazados de concreto simples. Determinação da absorção de água, teor de humidade e área líquida. Método de ensaio". Rio de Janeiro, ABNT, 1991 (in Portuguese).

[5] NBR 1228 – "Cálculo de alvenaria estrutural de blocos vazados de concreto". ABNT, Jul./1989 (in Portuguese).

[6] NBR 13279 – "Argamassa. Determinação da resistência à compressão. Método de ensaio. Rio de Janeiro, ABNT, 1995 (in Portuguese).

[7] NBR 6136 – "Blocos vazados de concreto simples para alvenaria com função estrutural". Especificação, Rio Janeiro, ABNT, 1982 (in Portuguese).

[8] NBR 7171 – "Blocos cerâmicos para alvenaria. Método de ensaio". Rio de Janeiro, ABNT, Nov. 1992 (in Portuguese).

[9] NBR 7184 – "Blocos vazados de concreto simples para alvenaria. Determinação da resistência à compressão. Método de ensaio". Rio de Janeiro, ABNT, 1982 (in Portuguese).

[10] NBR 7217 – "Determinação da composição granulométrica dos agregados. Método de ensaio". Rio de Janeiro, ABNT, 1982 (in Portuguese).

[11] NBR 7218 – "Determinação do teor de argila em torrões nos agregados. Método de ensaio". Rio de Janeiro, ABNT, 1982 (in Portuguese).

[12] NBR 7219 – "Determinação do teor de material pulverulento nos agregados. Método de ensaio" Rio de Janeiro, ABNT, 1987 (in Portuguese).

[13] NBR 7220 – "Avaliação das impurezas orgânicas das areias para concreto. Método de ensaio". Rio de Janeiro, ABNT, 1982 (in Portuguese).

[14] NBR 7222 – "Argamassas e concretos. Determinação da resistência à tracção por compressão diametral de corpos de prova cilíndricos". Rio de Janeiro, ABNT, 1987 (in Portuguese).

[15] NBR 7251 – "Agregado em estado solto. Determinação da massa unitária. Método de ensaio". Rio de Janeiro, ABNT, 1982 (in Portuguese).

[16] NBR 8042 – Blocos cerâmicos para alvenaria – Formas e dimensões – Método de ensaio. Rio de Janeiro, ABNT, Nov. 1992 (in Portuguese).

[17] NBR 9776 – "Agregados. Determinação da massa específica de agregados miúdos por meio de frasco Chapman. Método de ensaio". Rio de Janeiro, ABNT, 1982 (in Portuguese).

[18] ROMAGNA, R. HELEO – "Resistência à compressão de prismas de blocos de concreto grauteados e não-grauteados". Dissertação de mestrado, Curso de Pós-Graduação em Engenharia Civil, UFSC, 2000 (in Portuguese).

[19] ROMAN, H. R. – "Determinação das características físicas e análise estatística da capacidade resistente de tijolos cerâmicos maciços". Dissertação de mestrado, Curso de pós-graduação em Engenharia Civil da UFRGS, Porto Alegre, Outubro 1983 (in Portuguese).

[20] ROMAN, H. R. – "Alvenaria Estrutural: Vantagens, Teoria e Perspectivas". 10° Encontro Nacional da Construção (ENCO), Gramado, RS, Novembro 1990 (in Portuguese).

[21] ROMAN, H. R. – "Argamassas de Assentamento para Alvenarias – III Simpósio de Desempenho de Materiais e Componentes de Construção Civil, Florianópolis, Outubro de 1991 (in Portuguese).

[22] ROMAN, H. R. – "Características Físicas e Mecânicas que Devem Apresentar os Tijolos e Blocos Cerâmicos para Alvenarias de Vedação e Estrutural". III Simpósio de Desempenho de Materiais e Componentes de Construção Civil, Florianópolis, Outubro de 1991 (in Portuguese).

CHAPTER 3

Typical masonry wall enclosures in China

Xianglin Gu

Professor
Tongji University
Shanghai
China

Bin Peng

Lecturer
University of
Shanghai ST
Shanghai
China

Xiang Li

Lecturer
Tongji University
Shanghai
China

SUMMARY

Masonry has been used for a very long time in China, and it still be widely used for many new structures. Materials used for masonry walls, types and characteristics of masonry walls, and applications of masonry walls in different kinds of structures, are described in this paper. The probabilistic theory based design method of masonry structures according to the current Chinese code and trends of development of masonry structures are also introduced.

1   APPLICATION OF MASONRY STRUCTURES IN CHINA

Masonry structures have experienced a long period of development in China. At the beginning, crude stone was used to build dwellings, and then worked stone was used to make different kinds of structures. Anji bridge in Hebei province is one of the most famous examples of these kind of structures. Although brick was used later than stone, it was adopted as the major material of the Great Wall, one of the most famous masterpieces of civil engineering around the world (Figure 1).

(a) Anji bridge in Hebei province

(b) The great wall

Figure 1.   Famous masonry structures of antique China.

(a) Masonry building of 12 stories
in Chongqing

(b) Reinforced masonry building
of 18 stories in Shanghai

Figure 2.   The tallest masonry buildings of modern China.

After the appearance of Portland cement in 1824, concrete block came into use in 1882. In Shanghai of China, 25 multi-storey dwellings with load bearing walls consist of concrete blocks were built in 1923.

After 1949, masonry structures have been booming in China. With the appearance of new materials, technologies and structural systems, the field of application has been extended, meanwhile the calculation theory and design method have been developed.

In 1996, yield of fired clay brick in China was 630 billion pieces, and it was equal to the sum of other countries in the world. In the same year, 67% of the new city dwellings were constructed as the mixed structure of brick masonry and concrete, and masonry walls were used in 90% of all buildings.

In order to save the non-recycle land resources and achieve the goal of sustainable development, a lot of perforated bricks have been produced since 1960 based on the assimilation of advanced technologies abroad and the circumstances of China.

Blocks made up of concrete, light aggregate concrete, air concrete or waste materials are also used widely in recent years, and the yield of building blocks had reached $7000\,m^3$ in 2004.

With the development of the theory for reinforced masonry structures, many economic and practical tall masonry buildings have been constructed all around China. Number of stories reached 12 and 18 for masonry and reinforced masonry structures, respectively (Figure 2). And experiences of constructing masonry structures in seismic area have been accumulated (1).

2   MATERIALS FOR MASONRY

2.1   *General*

In China, the masonry includes plain masonry, reinforced brick masonry and reinforced block masonry. The plain masonry can be sorted into 3 classes according to the material used: brick masonry, block masonry and stone masonry.

According to style of reinforcement, there are 3 classes of reinforced brick masonry: reinforced masonry members with horizontal steel grids, mixed masonry members with reinforced concrete cover or reinforced mortar cover, mixed masonry walls with structural concrete columns.

To resist vertical and horizontal loads applied on reinforced block masonry buildings, vertical and horizontal steel bars are usually planted in the central hollows of small concrete blocks, and then the hollows are filled with grout. From this point of view, steel bars and grout are also materials of masonry structures, besides brick, block, stone and mortar.

2.2   *Bricks*

A brick is solid or perforated and is made up of fired clay, shale, gangue, or fly-ash, whose size is stipulated by the code (Figures 3 and 4).

(a) Fired solid bricks   (b) Fired perforated bricks

Figure 3.   Fired bricks.

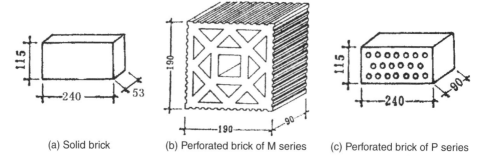

(a) Solid brick   (b) Perforated brick of M series   (c) Perforated brick of P series

Figure 4.   Size of fired bricks.

Bricks are defined as fired clay bricks, fired shale bricks, fired gangue bricks, and fired fly-ash-lime bricks respectively, if their ratio of hollow is less than 25%. If the hollow ratio is more than 25% and the brick is used for load bearing members, they are defined as perforated bricks which include M series and P series.

Assuring load capacity the structure, the clay resources and mortar can be saved, the weight and the cost can be decreased by using perforated bricks. Meanwhile, construction efficiency and seismic performance of the structure can be improved. With these advantages, the perforated bricks are used widely now.

According to their compressive and rapture strength determined through experiments, the fired solid and perforated bricks are classified into 5 grades, MU30, MU25, MU20, MU15 and MU10. Standard size of the fired solid brick is 240 × 115 × 53 mm. The typical size of the perforated brick is 240 × 115 × 90 mm for P series and 190 × 190 × 90 mm for M series.

a)   Gangue bricks
Gangue is one of the waste materials formed during producing and processing of coal. Volume of gangue formed each year is around 10–15% of the yield of coal in China. The deposition of gangue has got to 3 billion tons, and about 1500 hills of gangue have been making. It has occupied great deal of lands, also polluted underground water resources and the environment through the SO_2 emitted.

During the period of 1995–2000, technology of making gangue brick had been developed based on the assimilation of advanced technologies abroad.

Until the end of 2001, 8 production lines with yield around 30 million to 60 million pieces per year had come into use, and the increment of the production was 400 million pieces per year. 120 factories of gangue brick had been established in key coal mines of China, 2 billion pieces of standard bricks were produced per year. In the past 5 years, 20 thousand mus of lands have been saved by replacing the clay brick with the gangue brick (2).

Besides the gangue brick, there are still other kinds of bricks made up of industrial waste materials, including ferro-silicon slag, sulfuric acid slag, boiler slag, mine crumb, carbide slag and so on.

b)   Sand-lime bricks and fly-ash-lime bricks
Sand-lime bricks are made up of lime and sand mostly (Figure 5). Fly-ash-lime bricks are made up of fly-ash, lime, gypsum and other aggregates. They are solid

Figure 5. Sand-lime bricks.

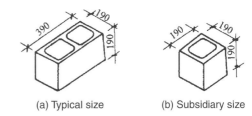

(a) Typical size   (b) Subsidiary size

Figure 6. Size of the concrete hollow block.

bricks making through several steps, including material collection, forming, and steam curing (3). They are classified into several grades according to their experimental strength, including MU25, MU20, MU15, and MU10.

Considering the strength of fly-ash-lime, bricks may increase with the depth of carbonization, a factor reflecting natural carbonization should be taken into account to determine the strength of the brick, and the factor might be 1.15 if there is no reference information.

2.3 *Blocks*

The concrete block used widely is the concrete small hollow block made by casting concrete or light aggregate concrete on forming machine. Typical size of load bearing concrete block is 390 × 190 × 190, and the ratio of hollow is 25–50% (Figure 6). According to the compressive strength determined by their gloss section, concrete block is classified into 5 grades, MU20, MU15, MU10, MU7.5 and MU5.

Concrete block is small, light, and can be used conveniently. Because of these advantages, it can be widely used in all kind of new buildings, and has become a substitution of traditional bricks.

a) Fly-ash-lime small hollow blocks

The fly-ash-lime small hollow block is made by mixing fly-ash, cement, aggregate, and water together, and admixtures are also used under some circumstances.

The weight of fly-ash used should exceed 20% of the weight of all materials. The weight of the wall constructed by fly-ash-lime small hollow blocks is about 2/3 of which constructed by fired clay bricks, so the seismic performance of the building is enhanced.

Meanwhile, construction cost of the foundation could be decreased by 110%, and construction efficiency could be enhanced to 3–4 times, also, the consumption of mortar could be decreased to 40%.

Furthermore, usable area of the building is enlarged, and the usage coefficient is enhanced by 4–6%. The gross cost could be decreased by 3–10%. Adiabatic coefficient of the wall could be 0.346 m K/W, so effect of insulation could be enhanced by 30–50%, and energy consumption of the building could be decreased (4).

b)   Sand-lime hollow blocks

The sand-lime hollow block is made by adding ash of sand into the autoclaved lime or cement concrete. The strength of this kind of the block is around 7.5–10 MPa. To decrease the cost, industrial waste materials containing silicon or calcium, including steel slag, carbide slag, fly-ash, and stone crumb, can be used if the strength of the product can be ensured.

Comparing with concrete hollow blocks, the shrinkage of this kind of the block is smaller because of its special forming process, and it needs not to be cured for 28 days. So the production period is shortened, companying with the reduction of the deposing field, the turnover of production founds is quite fast.

To enhance the early strength of the product, cement must be added into the mixing. According to the request of early strength and forming technology, hybrid technology similar to which used in the manufacturing of autoclaved concrete productions must be used in the manufacturing of this kind of the block, i.e., to replace some of the limes by cements, and replace some of the sands by stone pieces.

Just like the autoclaved concrete productions, the autoclaved sand-lime production is a kind of autoclaved production of calcium silicate, and their strength is offered by the calcium silicate generated through autoclaving.

c)   Light aggregate concrete blocks

The aggregate used includes pumice, shale stone, waste fired clay brick (5), boracic mud (6), expanded perlite (7). To enhance the strength of the masonry, shells and hollows of blocks should be aligned respectively. And insulation mortar must not be used for laying this kind of the block (8).

d)   Insulation blocks

The insulation block is one kind of composite blocks, and it is made by pressing the concrete cover, insulation part, and the common block together (Figure 7). Industrial waste materials can be used to produce this kind of the block. Advantages of the block include good insulation effect, and convenience for construction.

The block is developed based on the boiler slag block, and appropriate mixing ratio of the concrete and the quality of mixing are key factors during the

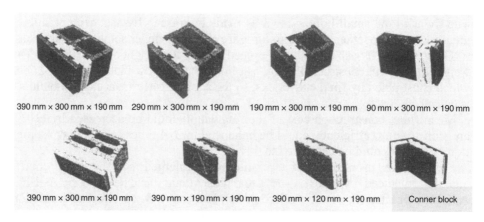

| 390 mm × 300 mm × 190 mm | 290 mm × 300 mm × 190 mm | 190 mm × 300 mm × 190 mm | 90 mm × 300 mm × 190 mm |

| 390 mm × 300 mm × 190 mm | 390 mm × 190 mm × 190 mm | 390 mm × 120 mm × 190 mm | Conner block |

Figure 7.   Insulation blocks.

Table 1. Types of stones.

	Type	Specification and size
Worked stone	Fine worked stone	Being worked finely, regularity of shape is ensured, denting of laying surface is less than 10 mm, width and height of the section are greater than 200 mm, and greater than 1/4 of its length.
	Moderate worked stone	Idem, but denting of laying surface is less than 15 mm
	Rough worked stone	Idem, but denting of laying surface is less than 20 mm
	Coarse worked stone	Being worked slightly, regularity of shape is ensured approximately, denting of laying surface is less than 25 mm, the height is greater than 200 mm.
	Coarse stone	The shape is irregularity, and thickness at the middle is greater than 200 mm.

producing. The grade of strength of this block is MU3.5, and the ratio of cement to boiler slag is 1 to 4.5.

2.4 *Stones*

In China, the stone used in building engineering is cut from natural rocks. According to the regularity of the shape of worked stone, stones are sorted into 5 types, fine worked stone, moderate worked stone, rough worked stone, coarse worked stone and coarse stone (Table 1).

The strength of the stone is rather high. According to the results of compressive test for 3 cubic samples, there are 7 grades of strength of the stone in the code: MU100, MU80, MU60, MU50, MU40, MU30, and MU20. Generally, the compressive strength and conductivity factor of heavy stone are high, also, the durability is good, but it is hard to work.

2.5 *Mortar and grout for concrete small hollow block*

Mortar is cementation material in masonry structures which consists of inorganic gelatinization, fine aggregate and water. Distribution of stress on the surfaces of blocks or bricks in masonry structures is evened by being plating with mortar. By using mortar, gaps between blocks are filled, air permeability of the wall is decreased, and insulation and frost resistance of the wall are enhanced.

Mortar can be classified into 5 classes according to its constitution: cement mortar, hybrid mortar, lime mortar, clay mortar and gypsum mortar. Besides, a new kind of mortar for concrete blocks consisting of cement, sand, blending material and admixture has being used in many regions. Adding into water properly, it can be mixed mechanically. So the working intensity at construction site is lightened, and the quality of construction is enhanced.

Common mortar is classified into 5 grades according to its compressive strength determined by the test of cubic sample of 28 days, and size of the sample is $70.7 \times 70.7 \times 70.7$ mm. The 5 grades are M15, M10, M7.5, M5 and M2.5. Mortar for concrete blocks is classified into 4 grades, Mb15, Mb10, Mb7.5 and Mb5.

Strength of mortar is 0 in determination of the strength of masonry structures in construction period, because the mortar has not gained its anticipating strength.

To ensure the strength of masonry structures, mortar should be of good workability, including plasticity and water retention. Mortar with good plasticity is easy to be distributed on the surface of blocks, so the mortar beds are homogeneous, satiation and density.

Water is not easy to be lost in mortar with good water retention during deposing, transportation and construction. Ingredients of mortar with good workability are not easy to separate from each other, so the mortar would not laminate or isolate, and the gelatinization material can work properly to ensure good joint of mortar and blocks, and then the strength of masonry is guaranteed.

To enhance the workability of the mortar and the quality of the construction, proper quantities of inorganic plasticizer, like lime paste, are usually added into the mortar. Considering there is no plasticizer in cement mortar, a coefficient is imposed when determining the strength of masonry constructed by this kind of mortar, and the coefficient is between 0.8–0.9 according to work conditions of the masonry.

In reinforced masonry structures, grout for concrete small hollow block is necessary. In reinforced concrete block masonry, the grout is made up of cement, aggregate, water, blending material and admixture, and can be mixed mechanically. The grout is used to fill the hollows of the block, and to form button stem.

To ensure the reinforcing steel bars are in the right position, the grout should be filled into the hole easily and not isolate, so fineness of the grout should be properly chose according to the size and the position of the filling hole.

To ensure the density of the grout, low-position filling and high-position filling should be used according to the position of the hollow. The grades of strength of the grout for concrete small hollow block are Cb20, Cb25, Cb30, Cb35, Cb40, Cb45, Cb50, Cb55, and Cb60.

2.6   *Wall plates*

The wall plate is a kind of big masonry member. Usually, the height of a wall plate is equal to the height of a story in the structure, and its width is equal to the width of a bay or the depth of a typical room. It is easy to realize industrialization and mechanization of building engineering, shorten the period of construction, and enhance productivity by applying wall plates.

Masonry wall plates include prefabricated brick plate, prefabricated block plate and vibrated brick plate. To produce wall plates, bricks/blocks and mortar are laid continuously using special equipment.

To produce vibrated brick plates, firstly a layer of high strength mortar is laid in the steel model and the thickness is 20–25 mm, then a layer of stagger jointed brick is laid laterally, and a layer of mortar is laid again, and then the specimen is vibrated, finally steam curing is performed. Mortar in the wall plate produced by the process is homogeneous and of good density, and the masonry is of good quality.

Comparing with 240 mm thick brick masonry walls, it can save 50% of bricks, decrease 30% of self weight, save 20–30% of workers, shorten 20% of construction period, and decrease 10–20% of cost to use 140 mm thick wall plates.

The wall plate can also be made by single material, like prefabricated concrete, miner slag concrete or in-situ cast concrete. Choice of the wall plate should be made according to the circumstances (9).

3 TYPES OF MASONRY WALLS

3.1 *Plain masonry walls*

a)   Brick masonry walls
Masonry walls can be constructed by different manners (Figure 8). Several kinds of masonries with different thicknesses can be constructed, including 240 mm, 370 mm, 490 mm, 620 mm, 740 mm, by using one brick, one brick and a half, two bricks, two bricks and a half, three bricks, respectively. And all kinds of bricks mentioned above can be used.

b)   Block masonry walls
Concrete single hollow blocks and light aggregate concrete blocks have been used widely in building systems.

Because partition of blocks is quiet difficult, using matching blocks and arranging them properly are key processes in the design, and it is different from the design of brick masonry walls. The regularity of arrangement should be ensured, and the number of types of blocks should be as less as possible.

To enhance the strength of the masonry walls, the number of straight joints should be reduced (Figure 9).

The typical size should be the first choice in construction. Blocks with thickness of 190 mm are often used to constructed load bearing walls if there is no request of insulation, and blocks are upside down in the wall.

c)   Stone masonry walls
This kind of masonry wall is constructed using stone with mortar or concrete, including worked stone masonry walls, coarse stone masonry walls and coarse stone concrete masonry walls (Figure 10).

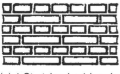 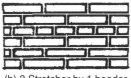 

(a) 1 Stretcher by 1 header   (b) 3 Stretcher by 1 header   (c) Quincunx header

Figure 8.   Bricklaying of masonry walls.

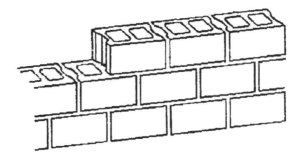

Figure 9.   Masonry wall of concrete small hollow blocks.

(a) Worked stone masonry wall   (b) Coarse stone masonry wall   (c) Coarse stone and
concrete masonry wall

Figure 10.   Stone masonry walls.

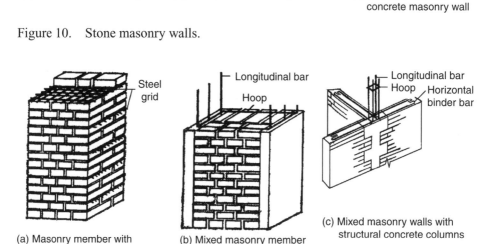

(a) Masonry member with   (b) Mixed masonry member   (c) Mixed masonry walls with
horizontal steel grid   structural concrete columns

Figure 11.   Reinforced brick masonry members.

3.2   *Reinforced brick masonry members*

The section size can be reduced and the strength can be enhanced through arranging reinforcing steel bars in brick masonry walls.

According to the arrangement of steel bars, reinforced brick masonry members can be sorted into 3 classes (Figure 11): reinforced masonry members with horizontal steel grids, mixed masonry members with reinforced concrete cover, or reinforced mortar cover, mixed masonry walls with structural concrete columns.

The axial compressive strength can be enhanced by arranging steel grids horizontally in masonry walls or columns. If the strength is not enough to bear eccentric

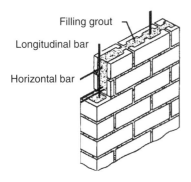
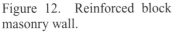
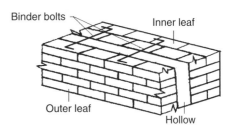

Figure 12.   Reinforced block masonry wall.

Figure 13.   Sandwich wall.

vertical loads, reinforced concrete cover or reinforced mortar cover can be used to enhance the strength of walls or columns.

At the cross joint of walls, end of walls and around the big hollows, structural concrete columns with spans less than 4000 mm can be used to form mixed masonry walls, and the strength of the wall can also be enhanced by this manner.

3.3   *Reinforced block masonry walls*

In order to improve the resistance of block masonry structures to axial or lateral loads, steel bars are arranged in the hollow of concrete block, or/and grout is filled into the hollow to strengthen bonding behavior between blocks and integrity of the structure (Figure 12). By these measures, high-rise masonry buildings could be constructed.

Mechanical performance, load bearing capacity, insulation of sound and heat, fire resistance of reinforced masonry walls are better than plain masonry walls. According to the requests of the design, hollows can be filled partially or entirely.

Besides the 5 types of load bearing masonry walls mentioned above, sandwich walls constructed by filling insulation materials into continuous hollows between the inner and outer leaf, and connecting the two leafs with medal bolts are used in northern China (Figure 13), and this kind of wall should be generalized because its thermo-technical performance is very good.

Traditional cavity walls had been widely used in southern China (Figure 14). But it is seldom used for new buildings because of its bad seismic performance and durability.

4   APPLICATIONS OF MASONRY WALLS IN DIFFERENT KINDS OF STRUCTURES

4.1   *Mixed structures of brick masonry and concrete*

It is the main structural type adopted in Chinese dwellings. Masonry is used to construct load bearing walls, and prefabricated or in-situ cast concrete slabs are used to make floors and roofs in the structure (Figure 15).

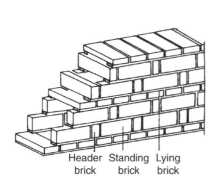

Header Standing Lying
brick   brick   brick

Figure 14.   Cavity wall.

Figure 15.   Mixed structure of brick masonry and concrete.

Figure 16.   Mixed structure of brick masonry and concrete with ring beams and structural concrete columns.

The walls are usually constructed by fired solid bricks and hybrid mortar, and perforated bricks are now widely used to replace solid bricks. The seismic performance of the structure is not very good, so ring beams and structural concrete columns are often used to strengthen its ductility (Figure 16).

Typical problem of the structure is cracking of walls during usage because of deformation due to change of temperature or shrinkage. Effective solutions are to control shape of the structure and arrange steel bars in temperature sensible parts of the walls.

4.2   *Mixed structures of brick masonry and timber*

They are the most representative structures of China, and most of the protected historical architectures are of this kind of structures. Load bearing walls including lintels are constructed by brick masonry, and the floors or roofs are made by timber in the structure (Figure 17).

Figure 17.   Mixed structure of brick masonry and timber.

Durability relating problems, including decency, frost-throw cycle, wet-dry cycle, corrosion of timbers, are typical problems of this kind of structures. Besides, the structure is vulnerable to disasters, including fire, earthquake, wind, and termite.

4.3   *Masonry structures supported by RC frames at the bottom*

In China, multi-function and big bays are the trends of development for masonry buildings. In masonry structures supported by RC frames at the bottom (Figure 18), big bays are achieved at the bottom story by using concrete frames, and dwellings are constructed by masonry at the upper stories.

The request of multi-function can be met. Load bearing features of members and the seismic behavior should be considered in the design of the structure, for example, there are interactions between the upper masonry walls and its supporting concrete beam of the frame (Figure 19).

(a) Elevation   (b) Supporting frame

Figure 18.   Masonry apartment building supported by RC frames.

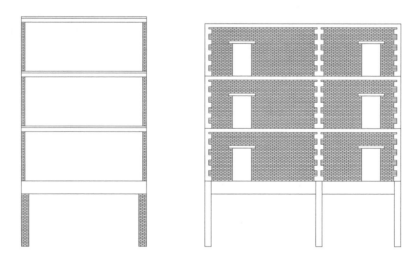

Figure 19.   Masonry walls and supporting concrete beam of frame.

4.4   *Framed structures*

Masonry is often used to constructed infill walls in frame structures.

4.5   *Bent structures*

Masonry is used to constructed enclosure walls in bent structures, which are often used for factory buildings (Figure 20).

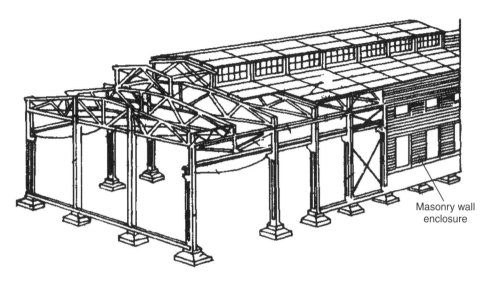

Masonry wall
enclosure

Figure 20. Masonry wall enclosure of a bent structure.

5 DESIGN METHOD OF MASONRY STRUCTURES

In China, design method of limit states based on probabilistic theory is adopted in the current Code for Design of Masonry Structures, numbered with GB50003–2001. Limit states of the structures include ultimate limit state and serviceability limit state, and the limit state equation is the base of design. The reliability of structures is measured by probability of failure or reliability index. The probability of failure should be limited in an acceptable range according to the code.

Because there are many factors affecting the safety of the structure and their regularities of distribution are different from each other, the probability method used is approximate. The First-Order Second-Moment Method (FOSM) considering the probability distributions of basic factors, like loads and materials, is used. Meanwhile, because it is not convenient to use probability index in calculation, formulates expressed by characteristic values and partial safety factors are still used for design.

In order to determined the formulates safely and economically, the target probability indexes are chose based on the failure modes of members, and the best partial safety factors matching different proportions of permanent loads and live loads are used to meet the requests of the design. And the error between probability index for members designed based on formulates of limit states and the probability index of the code is limited to the minimum (1).

According to the features of masonry structures, the requests from serviceability limit state can be meet under most circumstances if proper structural measures are used. And the design of ultimate limit states for a specific design reference period should be performed according to 3 grades of safety (Table 2).

Based on the engineering practices in China, to ensure the safety and economic of structures under the regular conditions of design, construction, and maintenance,

Table 2.   Safety grades of building structures.

Safety grades	Consequences of the failure	Types of building
Grade 1	Disaster	Important building
Grade 2	Severe	Common building
Grade 3	Mild	Minor building

the following formulate is used in the Unified Standard for Reliability Design of Building Structures and the Code for Design of Masonry Structures,

$$\gamma_0 S \leq R(\bullet) \tag{1}$$

$$S = \gamma_G S_{Gk} + \gamma_{Q1} S_{Q1k} + \sum \gamma_{Qi} \Psi_{Ci} S_{Qik} \tag{2}$$

$$R(\bullet) = R(\gamma_a, f, a_k, \ldots) \tag{3}$$

where:

- γ_0 is the coefficient of importance, and the value is 1.1 for the structures of grade 1 and design reference period is 50 years, 1.0 for the structures of grade 2 and design reference period is 50 years, 0.9 for the structures of grade 3 and design reference period is 1–5 years;
- $R(\bullet)$ is the function for resistance of members;
- γ_a is the adjust coefficient for design value of strength of masonry;
- f is the design value of strength of masonry;
- a_k is the characteristic value of geometry factor;
- S is the combination value of effect of actions, and could be expressed by axial force N, bending moment M, or shear force V generated by actions;
- γ_G and γ_{Qi} are the partial safety factor of permanent load and live load i, respectively;
- Ψ_{Ci} is the combination factor of live load i, the value is 0.7 generally, and is 0.9 for stake room, archival repository, stock room, and equipment room.

6   TRENDS OF DEVELOPMENT FOR MASONRY STRUCTURES

6.1   *Sustainable development*

Land saving, energy saving, and applicability are requests for masonry structures, so the industrial waste materials including fly-ash, coast slag, minor slag, boiler slag and endemic materials like lake mud, river mud, and shale rock should be the first choice of making blocks.

Because masonry constructed by fired clay brick has many disadvantages, like great energy consumption, wasting of lands, transportation burdensome, it can not meet the request of sustainable development, and is being eliminated gradually.

6.2   *Innovation of materials*

The hollow block with advantages of high strength, lightness, high performance should be used. The productivity can be enhanced, the weight can be reduced, the seismic performances and insulation can be improved, and the mechanism of load bearing is more proper by using this kind of blocks. The strength and seismic performances can also be enhanced by using mortar of high strength and good bonding properties.

6.3   *Innovation of technology, structural system, and design theory*

Seismic performance of reinforced masonry structures is much better than that of plain masonry structures. The reinforced masonry can be used to construct high-rise buildings, and it has been used more widely in China.

Working intensity can be lightened and the construction can be accelerated by using wall plates produced industrially and mechanically. It is also an advantage way to conduct pre stress on masonry walls. Proper arrangement of members in masonry structures, mechanism of resistance and failure are key aspects to be studied in the design theory.

7   REFERENCES

[1] Su, X. Z., Gu, X. L. – Design of masonry structures (in Chinese). Press of Tongji University, 2002.

[2] The gangue brick is becoming a replacement of fired clay brick (in Chinese). http://www.cqvip.com.

[3] Zhen, S. J., Liu, Y. J., Jia, Q. J. "The produce and design of flyash-lime brick" (in Chinese). Comprehensive utilization of fly-ash, (1), 2004: 53–55.

[4] Dong, W., Xu, H. S. "The developable fly-ash-lime small hollow block" (in Chinese). Brick and tilt, (6), 2005:41–42.

[5] Xin, Y. "Non-load-bearing hollow brick made up of waste clay brick" (in Chinese). Information of the industry of building material (1), 2003: 27.

[6] Xue, F. L."Production of expanded perlite block" (in Chinese). Brick and tilt, (2), 2005:36.

[7] Huang, L. H., Zhou, D. W. "Development of light aggregate block containing boron mud" (in Chinese). Building blocks and block masonry buildings (1), 2005: 33–34.

[8] Zhu, W. D., Zhou, Y. C., Cao, D. G. "Experiment research on masonry of light aggregate block containing pumice and shale rock" (in Chinese). Building blocks and block masonry buildings (1), 2005: 24–26, 52.

[9] Wang, C. Y., Zhi, D. W. "Produce of light aggregate composite and insulation block containing boiler slag" (in Chinese). Building blocks and block masonry buildings (2), 2005: 11–13.

[10] Shi, C. X. "Masonry structures" (in Chinese), 2nd edition. China building industry press, 1992.

CHAPTER 4

Typical masonry wall enclosures in France

J. D. MERLET

Engineer
Technical
Manager
CSTB
France

P. DELMOTTE

Engineer
Safety Engineering
Associated
Technologies
CSTB
France

SUMMARY

A short overview of current units, main constructive detailing and practical current solutions used in France may give a first opinion of the masonry place in our country.

On the other hand, some comments about simplified rules for design calculation and on future application of EUROCODE 6 allow to a better knowing of specificities of such buildings in France. At the end, some indications on possible future evolutions are given, underlined by references to the main results of research works already available.

1   INTRODUCTION

As agreed by CIB Commission W23 Wall Structures, opportunity to disseminate national overview on building masonry enclosure systems used in various countries has been considered of interest.

Because of its extreme variety from a country to another and even from some part to another one in the same country, masonry is probably the building technology which has got lot of problems in the field of European standardisation.

Such European overview may be a contribution to better understanding of such diversity: the purpose of this paper is to give a description of the French situation.

2   BUILDING SECTOR

France presents a great diversity of climatic situations from the north to the south and from east to west regions, with oceanic influence on the west half of the territory.

Masonry use is concentrated to small constructions and mainly to individual houses. It is also currently used for enclosure walls for collective dwellings, or for complete buildings the structure of which is made of reinforced concrete frame.

Recent general evolution in France from 10 years ago is disposition of building construction of minus to 5 storey height.

3 MASONRY UNITS USED IN FRANCE

Masonry units used in France probably more than in other countries in Europe made of various constituent materials but also various dimensions and internal structures.

We can find in France around all its regions from northern to southern all types of masonry units as described and defined in products standards established by CBN TC 125 in BN 771 series except for calcium silicate units which are not produced in France and so only slightly used.

In fact, majority of units are concrete and clay products; moreover mainly hollow units the holes of which may be pass through or not and vertically (concrete and clay products) or horizontally products (clay products only); such units allow to build lightweight masonry, the weight by square meter of which don't exceed around 150 kg/m^2 (because of very thin internal wall or shell of units). Such masonries are currently rendered almost on the external side.

The rest of masonry units used to are faced units (small clay units $6 \times 11 \times 22$ like Figure 2a) or large concrete units $20 \times 20 \times 50$ hollow blocks the shell of which are more than 25 mm thick, like Figure 1c or solid blocks made of aerated autoclaved concrete, very similar to products used in other European countries.

(a)   (b)   (c)

Figure 1.   Concrete units.

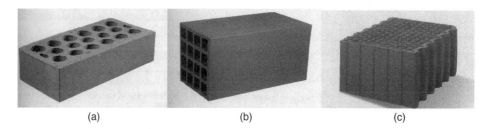

(a)   (b)   (c)

Figure 2.   Clay units.

4   DESIGN AND CONSTRUCTION DETAILING OF MASONRY WALLS

4.1   *General*

For other aim that stability such as thermal insulation, or merely to obtain relevant cohesiveness and avoid cracks in masonry, a majority of such constructions are comfortably design the cross section of loadbearing walls are largely up to the minimum should be according to the real loads supported.

It is very easy to reach such conclusion when we considered that most of masonry constructions in France are individual dwellings (or one family houses) or small constructions (less than two storeys height).

About cohesiveness, the most current use is by incorporating in the masonry a reinforced concrete ring beam and corner piles, for that reason and to allow in the same time an uniform background to renders, these concrete parts of work are covered on external part by thin units made of the same material than current masonry units: finally even for low construction works minimum thickness of wall is over 20 cm.

Cohesiveness and integrity are very important mainly about resistance to driving rain penetration of external walls: so renders currently made of mortar directly applied are very sensible to cracks in the background because such rendering mortar have no sufficient capacity to resist crack opening.

On the other hand cracks of hazardous lines give an aspect totally unacceptable for occupants. For the same reasons, this is not useful to execute on a smooth surface of rendered masonry wall some movement joints which don't be sufficiently spaced to be of an acceptable general aspect mainly for the "façade".

It is a fundamental difference between rendered masonry and faced masonry in which apparent joints give an easy possibility to hide the movement joints among current joints; and it is one of the main reason why masonry users in France payaccurately of attention on dimensional stability as important characteristic of masonry units.

Such use of rendered single walls in construction façades had been a considerable influence on design and construction detailing which are conducted mainly in function of the water tightness and air tightness of the whole façade walls.

4.2   *Water tightness function for masonry walls. Recall of some definitions related to water tightness*

A wall is watertight when it constitutes a relevant barrier to avoid that water don't progress from external face to its visible internal face; in addition such infiltrations in wall thickness should not affect significantly wall characteristics (such as thermal insulation) or its behaviour.

Wall tightness concerns both water tightness linked to raining water and also melt-ing snow, and air tightness very often linked to water tightness by water running by air phenomena.

It is to note that air tightness of façade concerns also thermal insulation problems (depredations, of course, but also occupant comfort: internal temperature, cold air perception, …).

The satisfaction to such requirements is to be fulfilled both in current parts of works and also in singular parts that are made by liaison of walls to adjacent works.

Air tightness problem are particularly accurate in case of double leaf walls, for instance walls covered by claddings on external side: in such case tightness is insured:

– in the internal wall plan in most of cases
– at the liaison of this internal wall to other works: window, stiffening walls, floors, electrical incorporation …

4.3 *Wall resistance to rain penetration*

In order to fulfil the water tightness function of a wall, it shall be adapted the resistance to rain penetration of that wall to the relevant severity of the site where the planned masonry construction will be build; the severity of the site can be characterized by:

– local situation of construction
– wind exposition
– general environment of site.

So main data to consider are:

– duration and frequency of sequences "wind-rain" mean and hygrometric ambiance on site of construction which conduct successive phases of wetness and dryness
– factors which regulate during rain risk and dynamic pressure of wind conduct severity of wetness sequence.

On the last topic, main relevant elements are:

– the situation of construction (from great urban centres until seaside)
– the tight up to the ground
– the presence or not of some protection against wind (mask effect).

In France in order to simplify for designer the same criteria have been adapted for choice of windows and also for thermal depredations, calculation of the part resulting of wind action.

4.4 *The answers brought by masonry walls*

The classification of wall in relation to their resistance to rain penetration is described here after and illustrated by the figures of Table 1.

The resistance to the rain penetration depends also on relevant devices incorporated in the wall, in the context of design for the various types of wall included in the classification:

Type I – single leaf wall
Such a wall don't contain internal capillarity barrier; it is not externally covered by a waterproof membrane: water penetration resistance is so directly dependant on water absorption of the wall itself, i.e. of its constituent materials and of course the thickness of the wall; it depends also on the good behaviour of the masonry wall and mainly the lack of cracks.

Table 1.   Classification of masonry walls on their rain penetration resistance.

Type II – wall solution
A first solution currently used to stop water penetration is to introduce inside the wall
and continuously a capillarity cut – i.e. air leaf or insulating board: in such a case efficacy is still dependant on the good behaviour of masonry (integrity of wall) but damp
proof course is sufficient to stop the progress of water penetration to internal side only
if its quantity is so small that it can be absorbed by unsaturated zones of masonry.

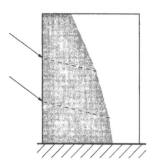

Figure 3.   Type I wall solution.

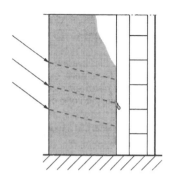

Figure 4.   Type II wall solution.

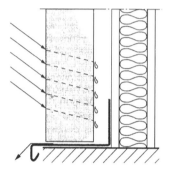

Figure 5.   Type III wall solution.

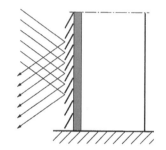

Figure 6.   Type IV wall solution.

Type III – wall solution

If severity of site is too high and so water penetration quantity in the masonry is too large to allow its absorption, by gravity action a part of that water may be conducted to the foot of air layer: in such a case it should be put at this place a damp proof course to collect water and reject it to external side through release which complete relevant disposal.

Type IV – wall solution

Absorption capacity is as much great as the capillarity cut is placed near internal side of masonry wall, but the most efficient way to avoid rain penetration by too high quantity when construction are very exposed, is clearly to stop rain before external side of masonry by mean of waterproof cladding system: such solution defined the type IV wall.

So, from type I to type IV resistance to rain penetration is improved (see Table 2 which gives some examples of various solutions).

Type I – Single leaf wall
Efficacy depends on:

– the thickness of the wall
– the capillarity of constituent materials of wall (masonry units, masonry mortar)
– and if it is the case mortar of rendering applied on external side of masonry.

Table 2.   Examples of types of walls showing improvement of rain penetration resistance from Type I to Type IV.

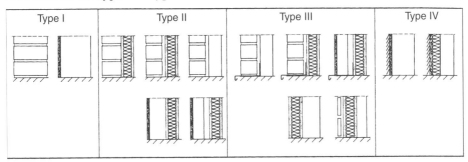

Type II – Double leaf wall
Efficacy depends on:

– the thickness (e0 or e1) of external wall
– the space between the two walls which constitutes a capillarity cut.

Type III – Double leaf wall with bottom foot disposal
Same constitution as type II but a damp proof course put at the bottom of wall which collect water passing through external wall and reject it on external side.

Type IV – Wall covered by cladding on external side
Cladding avoids rain penetration into the masonry wall by stopping it before external face.

5   PRACTICAL CURRENT SOLUTIONS IN FRANCE

The two most current types of walls used in France are:

1. Single load bearing wall made of hollow units (concrete or clay units) rendered on external side and completed by thermal insulated boards on internal side (thermal insulated panel plus masonry partition or laminated insulating plasterboard (see figure 4): such walls correspond to Type II.
2. Single load bearing wall made of high insulating masonry units: multi – alveolar clay units or autoclaved aerated concrete solid units; walls are rendered on external side and plastered on internal side: such walls correspond to Type I.

These two types of masonry walls are the most competitive solutions and correspond from external aspect point of view – by mean of render mortars using various colours or various funding coats – to the most current demand from the final occupant.

Such situation is the main reason of appearance on the market of an important industry of masonry and rendering mortars largely developed during the last twenty years: the best example of which are the one coat mortars which in these applications have for a major part substitute the use of in site mortar.

But there is also regionally mainly in the North of France and around Paris a current use of facing clay products (small units 6 × 11 × 22 cm).

The largely hollow units presents a similar internal structure with walls and shells of low thickness, which induces mechanical behaviour of a brittle failure characteristic from punching mortar in case of vertical perforations or buckling of walls in case of horizontal perforations.

This specific brittle behaviour is attenuated because of use of high resistance mortar in all case higher than usual other part in Europe, but it is one of the main reasons of generally good result (or absence of problems).

Masonry construction, the cross section of which are highly dimensioned and use of associated mortar also of higher resistance than strictly necessary by calculation rules, are both a specific aspects of current masonry use in France.

6   EUROCODE 6 (ENV1996.1.1) AND ITS APPLICATION IN FRANCE

We underline above some difficulties for a common application in France of Eurocode 6 (ENV1996.1.1).

As already state above; masonry is largely used in France for an application in small constructions, a majority of them are built with horizontally perforated clay units or hollow concrete units, to constitute what is commonly named lightweight masonry.

In view of design users applies currently simplified calculation rules which are part of French masonry code (DTU 20.1) relevant for such constructions.

On the contrary it should be difficult to apply Eurocode 6 particularly its first part (ENV1996.1.1); such statement should probably not so true about part 3 (PrEN 1996-3 "Simplified Calculation Methods") which are quite similar to such French code because it have been written as recipe more than design code model.

The main difficulties could find their origin in the fact that, as described above, current masonry in France, but also in Southern of Europe are made of hollow units.

Masonry constitutes in fact a very various and heterogeneous entity for which it should be difficult to find one only formula, or model to cover all types of masonry: so it would be necessary that such a formula give a relevant representation of the relation between the strength of that composite material which is the masonry and the strength of all its elementary constituent materials.

That observation have been shown as an evidence by an experimental series of tests including various types of masonry units used in France in the previous tests which have allowed to propose some guidance for more appropriate models (see cf *Traité de Physique du Bâtiment T.2 Mécanique des ouvrages Chapitre D4).*

That campaign of tests has also demonstrated that a behaviour law of parabola rectangle type by analogy to those used to reinforced concrete is not appropriate to many types of masonry and particularly those based on units of the internal structure; such masonry for that reason have a behaviour of a brittle type and from security point of view it is what is the most interesting in France.

It can be led a basic divergence about approach to the formulation of relevant model: on one hand by analytic approach, on other hand by empirical approach, based on heavy experimentation and numerous tests to determine for each masonry group, the parametric values to be introduced in one unique formula.

This last approach has two disadvantages:

– It needs as such above a preliminary experimentation very heavy which cannot be differently than evolutive, function of progress of experimentation that make difficult discussions of rules and attempt of consensus.
– It may induce reduction or penalisation as results in application of code or on the contrary be dangerous by unjustified grouping or insufficient definition of the limits of application for the empirical model so proposed.

7   STRUCTURAL DESIGN BY CALCULATION – SIMPLIFIED RULES APPLIED IN FRANCE (DTU 20.1)

7.1   *Masonry strength*

It is evaluated under action of vertical forces uniformly applied in the plan of masonry units strength characterised by standardised strength.

Such notion contain in fact various concepts: it means in reality the unit strength as measured in conformity to the relevant standard; until now one standard expresses relevant requirement as 5% fractile minimum strength, another gives a limit according to the mean value, associated to a minimum value applied to individual values obtained by testing. Such situation should be solved by available EN's.

But for this time, masonry strength value, represented by admissible strain (C) is then obtain by dividing the standardised strength value of masonry wall by a global coefficient (N): at this time there is one coefficient N specific for each unit formally.

Such coefficient which is inclusively evacuated contains as well as differences resulting from unit strength definition above, the consequences issued also respectively:

– from execution, uncertainties in brickwork, mortar used
– from slenderness ratio allowing to pass from masonry as material to masonry as work, up to 15 value – from 15 to 30, a supplementary reduction is applied to N (according to growth slenderness) and up to 30 we consider that it's no more in the case of current constructions and a particular study is required.

The eccentric input is inclusively evaluated by merely distinguish two cases:

– one for internal walls with centred loading
– another for external walls with eccentric loading.

About this second case of loading it is required a bearing length more or equal to 2/3 of load bearing wall thickness with minimum value around 20 cm in case of load bearing external walls.

7.2   *Stress evaluation*

This evaluation is limited to vertical forces resulting from gravity action (permanent loads, loads and snow …) and only horizontal forces resulting from wind action upon facades.

About evaluation of vertical loads acting on walls, it should be applied degressive low stated in NF P 06 001, taken in account that floor elements are discontinuously borne on walls.

About evaluation of strength to wind perpendicular to façade, wall panel is compared to a plate merely borne on its sides.

It is neither taken in account stress resulting from shrinkages and dilatations, for which it should be considered that fulfilment of construction detailing, allowing for neglecting them.

Nor those resulting from contribution of masonry wall to stiffening the construction, resulting of seismic actions or exceptional actions (impact, explosions, …).

7.3   *Calculation for validation*

Calculation hypothesis is that of uniform distribution for strains, except those resulting from loads of floor elements, lintels, etc., placed immediately above relevant horizontal section.

Figure 7(a) to (d) show the case of a slab or a beam perpendicular to mean plan of the wall given a triangular or trapezoidal repartition bearing in mind that the bearing length is also limited to the slab thickness or the length of beam.

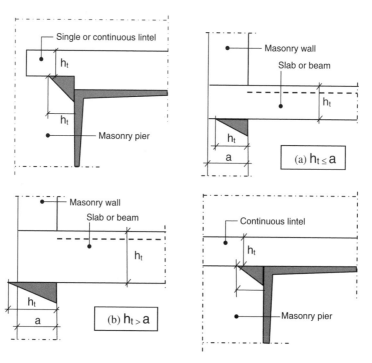

Figure 7.   Bearing stress repartition for a slab or a beam perpendicular to mid plan of the wall.

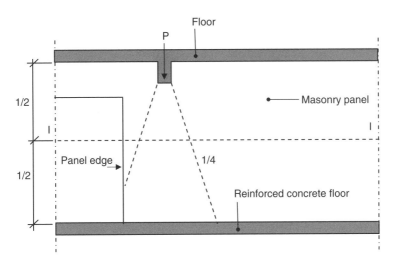

Figure 8.   Repartition angle in case of concentrated load.

Calculation consists in verification of strain:

– First, on current part (commonly at half height of the wall)
– and, also, on singular points such as piers, lintel bearings, beams, floor slabs.

About concentrated loads, it is considered that they are uniformly reported inside the zone limited by the two straights started from the application point of the load and inclined of 1/4 on the vertical direction (see Figure 8); such verifications are:

(a) under only vertical load actions, compressive stress should be at the most equal to admissible stress.
(b) under action of both the vertical loads and of wind perpendicular to façade, compressive stress should not be up to 9/8 of admissible stress.

8   TRENDS OF EVOLUTION

8.1   *Introduction*

Technical evolution depends on various combined factors including economical requirements, working customs, national regulations evolution etc…
We notice that very advanced concepts, (like both faced masonries assembled without mortar and with integrated thermal insulation for example), differ too much from working customs and generally need specific and embarrassing precautions. That's why the most relevant evolutions are those using systems which do not differ too much from classical techniques.

8.2   *Improvement of productivity and facility of assembly*

The main evolutions consist in the development of:

– mortar less assembled masonry, generally constituted by shuttering hollow blocks, completely or partially filled on site by premix concrete or mortar;

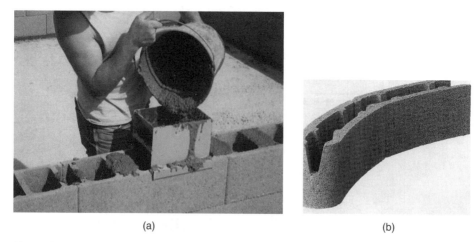

(a)   (b)

Figure 9.   Shuttering units assembled without mortar joints.

(a)   (b)   (c)

Figure 10.   (a) Masonry bedded with thin mortar joints; (b), (c) New tools for applying thin joints.

– masonries assembled with thin joints, bedded with special tools like spatulas or mortar rollers;
– larger unit dimensions so as to put fewer blocks per square meter, without exceeding however 20 to 25 kg weight per unit. We notice that systems using bigger dimensions units, requiring cranes of lifting machines, have difficulties to be developed;
– systems including special units, which avoid cutting current units with circular saws on site;
– masonry assembled with unfilled vertical joints, especially in seismic zones (see farther).

8.3   *Improvement of working conditions and organisation on site*

The main evolutions in this domain are:

– use of mortar less masonry techniques described above: they generally avoid noisy tools (concrete mixers, circular saws);

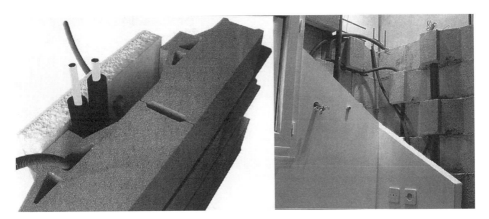

Figure 11.   Masonry systems incorporating technical networks.

– advent of special masonry walls systems which permit to incorporate electric, hydraulic and other technical networks before installing the insulating dry lining inside;
– advent of precast basement walls systems which allow to make the foundation operations less painful, allows to cast off oneself of weather conditions, and offer better linkage between basement and façade walls.

8.4   *Improvement of thermal performance*

This improvement is especially due to evolution of thermal regulations which now impose minimal values for thermal transmission coefficients. It concerns both thermal resistance and thermal inertia, with taking care to not however decrease mechanical performances.
   The main evolutions in this domain concern the development of:

– single loadbearing walls assembled with high insulating masonry units, which provide by themselves enough thermal resistance to satisfy thermal regulations, and offer in the same time a good thermal inertia. In order to get sufficient thermal values, these units generally have an important thickness, a multi-hollowed internal shape, and are made with aerated clay or lightweight concrete made with perlite, pumice or expanded clay aggregates.
– low density aerated autoclaved concrete units (less than $400 \, kg/m^3$);
– masonry systems using mortars made with lightweight aggregates, or containing air voids obtained by specific additives mingled with mortar on site;

These thermal performances are more often certified and regularly controlled by a certification body.

8.5   *Diversity of façade wall aspects*

Facing and/or curved masonry walls are now often used in architectural designs, with concrete or clay units, assembled in a classical way: classical bedding to valorise diversity of aspects. Special curved units could be used to obtain these results (see Figure 13).

Figure 12.   "Monomur" clay walls.

Figure 13.   Aspects of facing masonry walls.

8.6   *Environmental and health aspects*

These aspects are now foreseen in new assessments. Many aspects are examined, like energy consumption, waste, pollution impact, chemical and radioactive emissions, fungous expansion etc…

8.7   *Advances about unfilled vertical joints in seismic zones*

French regulations actually require filled vertical joints in seismic zones. However, for practical and economical considerations, it is obvious that the vertical joints

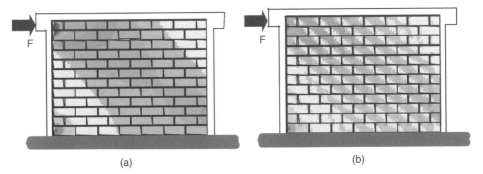

(a)   (b)

Figure 14.   Stress isovalues for walls with filled and unfilled vertical joints.

Figure 15.   Experimental configuration.

are not systematically filled while erecting the masonry walls, except for explicit requirements (such as fire exposure, water resistance, airborne sound insulation, etc.).

The aim of recent research was to evaluate the strength of masonry wind-braced walls made of clay or concrete units, with either filled or unfilled vertical joints, under a cyclic horizontal loading effect (see experimental configuration in Figure 15), applied on the top of the wall.

The test results show that the cracks follows the horizontal joints and are very dependent on the slenderness of the units, the compression strut following the diagonal of a half brick in the case of unfilled vertical joints (see Figure 14b), and following the wall diagonal in the case of filled vertical joints (see Figure 14a).

A simplified model based on the compression strut has been proposed. It is based on the mechanical and dimensional properties of the constitutive units on one hand, and on the wall overall dimensions on the other hand.

The results are in a good accordance with the experimental. These results could modify French regulations in the future, and allows taking into account the resistance of masonries with infilled vertical joint right now in new assessments.

9   REFERENCES

[1] « Flambement des murs en maçonnerie, élaboration d'un modèle pour murs non armés à section et matériaux quelconques ». Cahiers du CSTB n° 2397, livraison 307, Mars 1990, (P. Delmotte, JD Merlet, M. Pouschanchi) (in French).

[2] « Etude par simulation numérique du comportement des produits de maçonnerie et de leurs assemblages ». Cahiers du CSTB n° 2506, livraison 321, Juillet-août 1991 (P. Delmotte) (in French).

[3] « Maçonneries armées dans les joints horizontaux, règles professionnelles simplifiées ». Editions de l'union nationale de la maçonnerie, Juillet 1991, (in French).

[4] « Flambement des murs en maçonnerie. Élaboration d'un modèle pour sections de forme quelconque ». Cahiers du CSTB, livraison 325, n° 2540, Décembre 1991, 12 pages, (J. Lugez, P. Delmotte, M. Pouschanchi) (in French).

[5] « Résistance des maçonneries sous charges verticales, proposition d'un modèle simplifiéde calcul ». Cahiers du CSTB, livraison 326, n° 2553, Janvier-Février 1992, 11 pages (P. Delmotte, J. Lugez, J. D. Merlet) (in French).

[6] « Mathematical model applied to AAC masonry with thin joints. Advances in autoclaved aerated concrete ». Proceedings of the 3rd RILEM International Symposium on Autoclaved Aerated Concrete, Zurich 14–16 October 1992, pp. 251–257 edited by Folker H. Wittmann, Swiss Federal Institute of Technology, Zürich, (J. D. Merlet, P. Delmotte).

[7] "Les maçonneries à montage simplifié ». Cahiers du CSTB, livraison 335, n° 2623, Décembre 1992, 25 pages, (J. D. Merlet, P. Delmotte, B. Mesureur, D. Remy, J. L. Salagnac) (in French).

[8] « La modélisation de la maçonnerie armée par la méthode des éléments finis. Élaboration d'un modèle de calcul MOCA ». Cahiers du CSTB, livraison 337, n° 2639, Mars 1993, 22 pages (G. Mounajed) (in French).

[9] DTU 201 – « Parois et murs en maçonnerie de petits éléments ». CSTB, 1996 (in French).

[10] « Un outil pour la conception des ouvrages maçonnés ». BITMAC, les escaliers du progrès, Dossier Informatique et Construction, CSTB Magazine, n° 93, Avril 1996, pp. 9–12, (P. Delmotte, S. Wamba-Fosso) (in French).

[11] « Calcul des murs en maçonnnerie sous chargement hors plan ». Cahiers du CSTB, livraison 373, n° 2909, Octobre 1996, 22 pages, (L. Davenne, P. Delmotte, S. Wamba-Fosso) (in French).

[12] « Innovation et productivité dans le gros-œuvre. L'évolution du métier de maçon. Les conditions du succès: collaboration industrie, maître d'oeuvre, chantier; joints collés: blocs béton, briques de terre cuite à l'épreuve du chantier ». Dossier Maçonnerie, gros œuvre; CSTB Magazine, n° 99, novembre 1996, pp. 11–16; (P. Delmotte, B. Mouaci) (in French).

[13] « Fiabilité des murs en maçonnerie sous charge de vent, effets de la variabilité mécanique des joints et incidence de l'erreur de modélisation ». Cahiers du CSTB, livraison 392, n° 3065, Septembre 1998, 14 pages, (A. Sellier, A. Mebarki, P. Delmotte, C. El Hage) (in French).

[14] « Etude des murs de contreventement en maçonnerie d'éléments de terre cuite à joints verticaux secs ». Cahiers du CSTB, livraison 407, n° 3199, Mars 2000, 17 pages, (JI Cruz Diaz, A. Sellier, B Capra, P. Delmotte, P. Rivillon, A. Mebarki) (in French).

[15] « Modélisation simplifiée du comportement à rupture des murs ». Revue Française de Génie Civil, volume 5 n° 5/2001, Editions Hermès (J. Cruz Diaz, A. Sellier, B. Capra, P. Delmotte, P. Rivillon, A. Mebarki) (in French).

[16] « Murs de contreventement en maçonnerie de terre cuite. Approche expérimentale et modélisation du comportement à la rupture ». Cahiers du CSTB, livraison 416, n° 3310, Janvier-Février 2001, 16 pages, (J. Cruz Diaz, A. Sellier, B. Capra, P. Delmotte, P. Rivillon, A. Mebarki) (in French).

[17] « Fiabilité des murs de contreventement en maçonnerie. Calibration des coefficients partiels d'un modèle simplifié ». Revue Française de Génie Civil, Fiabilité des ouvrages de Génie Civil, Conception et maintenance volume 6 n° 3/2002, Editions Hermès (J. Cruz Diaz, A. Sellier, B. Capra, P. Delmotte, A. Mebarki) (in French).

[18] « Resistance of Masonry Infill Walls to Racking Loading: Simplified Model and Experimental Validation. Masonry International, volume 15, n° 3, 59–86 2002 (J. Cruz Diaz, A. Sellier, B. Capra, P. Delmotte, P. Rivillon, A. Mebarki).

CHAPTER 5

Typical masonry constructions in Germany

Wolfram
JÄGER

Professor
Dresden
University of
Technology
Germany

Peter
SCHÖPS

Associate Faculty
Dresden
University of
Technology
Germany

SUMMARY

This report should clarify the significance of masonry structures in Germany. It will be shown that masonry structures have a dominant position primarily in the residential construction sector. The historic development has been determined by the economic competition with other building materials and the increased demands by clients and the legislation. This led to a wide product range and a great variety of masonry construction types. However, research and development continues. More stringent requirements for thermal insulation and further reductions of the costs will lead to new inventions.

1 INTRODUCTION

Masonry has always played an important role in Germany, not only for protection against the elements but also for artistic reasons. The flexibility at the production level is a beneficial characteristic of masonry and contributes to its significance. A variety of building components are comprised of simply stacking stones with the aid of mortar. Therefore, it is possible to build straight or curved walls with arbitrary window and door openings and even vaults.

The task of the walls is therefore about:

- creation and definition of spaces;
- transfer of loads and forces, thus acting as structural building components;
- protection of the occupants against moisture, noise and fire;
- minimisation of energy consumption.

Today, other materials, like concrete or steel, are better suited to perform some of the above noted functions. However, as a result of ever changing developments, the potential uses, to build with masonry are also continuously increasing.

2   BUILDING SECTOR CATEGORISATION

2.1   *Introduction*

Buildings can be classified based on a variety of categories. Aside from the basic classifications such as building age, new or old, differentiation based on the function is especially common. Generally, the categories or types used are: residential, industrial, commercial, office buildings, public buildings etc. It already becomes evident in the former statement that differentiation according to the type of ownership of the structures, whether public or private, is also possible.

Another possible differentiation for the categorisation is the subdivision based on the building materials. The main types are: masonry construction, concrete construction, timber engineering and steel/glass constructions. Other potential categorisations which are possible include design typologies, structural systems or building height.

2.2   *Importance of the building sector*

The significance of the various building types and therefore also the share of the market which each type comprises, changes over the years. It is directly dependent on the technological advances and on the current aesthetic preferences of the society.

Cost effectiveness plays another important role. The goal of the designer and builder is to create a building with the proper functionalities and with as low a budget as possible. Whereas in former times material costs used to be the driving factor, labour costs are now the decisive factor.

Until modern times, essential buildings such as castles, cloisters or churches were primarily constructed of masonry, either from natural stones or bricks. In contrast, the majority of the residential buildings were limited to half-timber or timber constructions. Disastrous fires finally led to the understanding that buildings made of stone have several advantages regarding safety and durability, which over the life cycle is more cost effective.

Figure 1.   Example of half-timber buildings.

Figure 2.   Frauenkirche in Dresden before   Figure 3.   Brick-facing church.
destruction.

Within the last century the architectural style of masonry construction had to give way to the "more modern methods" of steel, glass and concrete.

These materials provided architects and engineers with new possibilities for design and construction. This primarily affected industrial and public buildings. However, changes also occurred in the residential sector.

In the 1950's, structures started to be built predominantly out of prefabricated concrete elements. In the Eastern part of Germany (formerly the GDR) the so-called "panel construction settlements" have developed. This architectural style became evermore common practice until the German Unification in 1989. However, the design possibilities are limited and production orders with just a few pieces usually are not economical. Thus the use of masonry especially for building houses increased again.

An important area for the application of masonry was and still is residential construction. These buildings are predominantly built of masonry.

2.3   *Structural systems for buildings*

The architect can choose from a variety of possible construction types when designing a building. It is possible to change the building material and thus construction type in one and the same building.

Apartment house of autoclaved
aerated concrete units (YTONG)

Apartment house of calcium silicate units [4]

Apartment house of calcium silicate units

One-family house of chipped wood insulating
masonry units for concrete infill

Two-family house of clay units [5]

Two-family house of clay units

Figure 4.   Examples of residential houses, still without plaster.

The structural framework is one example. It can consist of columns, floors and ceilings. Generally speaking the structural elements are built out of reinforced concrete. The infilling partition can, of course, also be built in masonry. Steel structural systems are also used for high strains.

The use of wood for load-bearing building components is mainly confined to traditional roof constructions. The load-bearing framework of half-timber buildings consists of timber columns and beams, which are stiffened with diagonal or cross members. The walls are then infilled, either with masonry or similar materials.

Another method is timber panel construction. Here a load-bearing timber construction similar to half-timber buildings is used. Insulating materials in-filled between the various timber elements. Pressboards are affixed, inside and outside. The facing layer attached on the outside, guarantees protection against the elements and consists for example of clinkers (facing bricks).

The construction principle to build masonry walls, reinforced concrete ceilings and a timber roof structure frequently with a saddle roof shape, is commonly used in Germany.

Brick ceilings in combination with a poured in place concrete layer are sometimes substituted in practice in place of the reinforced concrete ceilings.

2.4   *Systems used for housing, commercial, etc.*

Today, single-family houses and multiple dwelling units are built with the architectural style that uses masonry, reinforced concrete ceilings and saddle roofs. Further, adapting the floor plans to the special requirements of the occupants is possible without a high design effort. Beyond that, the Summer and Winter thermal heat insulation is guaranteed without the need for any additional HVAC equipment. This leads to economical buildings.

Modular homes offered, for example, in home-improvement stores and shopping centres are economically priced and generally built with the timber-panel style. In these cases, the walls do not have to transfer heavy loads. Therefore construction of this type of building is very fast. However, the living climate is less favourable because of the low heat retention capacity of dry construction walls.

For public and government buildings the scope of the design is bigger and thus also the choice of the construction type. In this type of buildings it is the aim to build in such a manner as to allow a great deal of transparency. Therefore a steel glass construction is often used. Structurally more demanding buildings consist of reinforced concrete. Masonry only serves as a creative element for these buildings.

Office buildings are mainly built with a skeleton construction. Thus changes in the floor plans or changes of the building-use can easily be accomplished solely with alterations of the interior partitions. Masonry can be used as a façade element, or inside for highly frequented areas or as fire-rated walls.

Industrial buildings such as factory buildings are built from prefabricated reinforced concrete elements or from steel frames in the case of larger spans. Older industrial plants are often impressive brick buildings with fair-faced masonry.

2.5   *Some statistics on use and relevance*

The percentage of masonry buildings is the largest in residential construction as already noted in the previous chapter.

By comparing solely the number of issued building permits of each construction type a somewhat distorted representation, relative to the overall construction activities in Germany, is produced because the size of the projects are not considered.

When considered, masonry and concrete construction types comprise an even larger position of the market relative to other construction types, as opposed to when looking only at the quantity of issued building permits. This indicates that masonry and concrete buildings comparatively, comprise a greater quantity of useable living space than the buildings of the other construction types.

Interestingly enough, the construction costs of timber buildings, percentage wise, is higher than the percentage of building volume constructed. This confirms the assumption that this construction type does not contribute, at least statistically speaking, to building with cost-savings.

For masonry constructions the percentages are almost the same for the two considered factors, i.e. cost and building volume.

The general decline in the building industry also affects the sales of the masonry units industry.

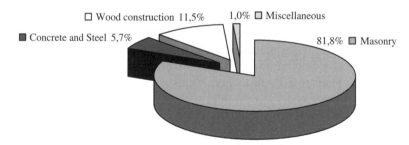

Figure 5.   Percentages of the issued building permits for the various construction types. (Source: Federal Statistical Office, statistic on building activities, building permits 2000)

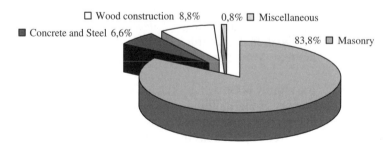

Figure 6.   Percentages of the various construction types of building volume in residential applications.
(Source: Federal Statistical Office, statistic on building activities)

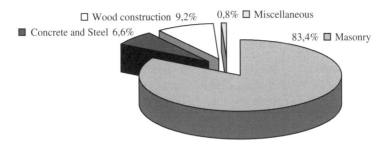

Figure 7.   Percentages of the costs of the various construction types relative to the total construction costs in residential buildings (41.567 billion €).

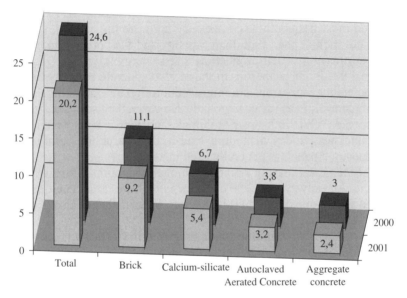

Figure 8.   Market share of the masonry units industry [m3].
(Source: DGfM, based on member supplied data)

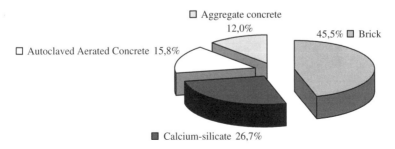

Figure 9.   Market share of the masonry units industry [%].
(Source: DGfM, based on member supplied data)

3   MOST COMMON ENCLOSURE WALL SYSTEMS

3.1   *Masonry material*

a)   Masonry units
In Germany, masonry units are divided into four major categories. These are clay bricks, calcium silicate units, autoclave aerated concrete masonry units, and lightweight pre-cast concrete or pre-cast concrete masonry units. Additionally, there are natural stones such as sandstones.

As illustrated in Figures 8 and 9, bricks are the most frequently used masonry units that are used. In the past, mainly smaller bricks were used. Then larger bricks with vertical holes were developed. The so-called clinkers are used for façade panelling or fair-faced masonry. These are bricks, which were fired at a higher temperature and therefore have a closed surface for better weather resistance.

The second most frequently used masonry units are the calcium silicate units. The advantages include their high load capacity and the potential to manufacture large units. Furthermore, their production is less energy intensive.

The thermal insulation properties of the aerated concrete masonry units and the pre-cast concrete masonry units are improved through air-entraining agents.

The use of natural stones varies regionally. However, the use of sandstone is relatively widespread. While in historic buildings sandstone ashlars were used for the entire wall thickness, today a modern load-bearing structure combined with a veneer of sandstone panels is generally used instead.

In some of the newer masonry units additional insulation is already integrated through the use of polystyrene. The masonry units have larger cavities which are then in-filled with expanding liquid insulation. Special shapes also get produced, e.g. window jambs, ring beams or lintels.

Today the quality of the masonry units produced in Germany is very high. Due to the precision of the masonry units, the mortar thickness used to overcome dimensional inaccuracies can continually be decreased.

b)   Mortars
The three traditional mortar types – cement lime-, cement- and lime mortar – are still commonly used in Germany. Beside the noted binders, the mortar also

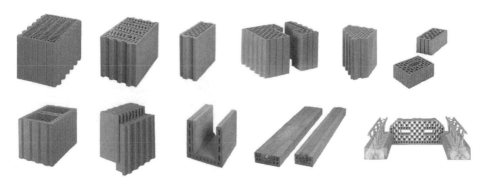

Figure 10.   Examples for bricks [5].

contains differing portions of sand. These mortars can be mixed at the construction site from a variety of components. However more frequently, they are delivered pre-mixed and only water needs to be added on site.

Lightweight mortar and thin-bed mortar are other options beyond the traditional mortars. Higher masonry strengths can be achieved by using thin-bed mortars. Lightweight mortars are suitable for constructions consisting of the masonry units that have high resistance to heat transmission. This allows thermal bridges to be avoided in the areas of the mortar.

More and more frequently, masonry units in conjunction with a specific mortar are defined as a system. Consistent quality of the masonry work is ensured due to the fact that the masonry units can only be used in combination with the specified mortar, which must be delivered by the same manufacturer.

c) Reinforcements and wall ties

The use of reinforced masonry is not common in Germany. However if reinforcement is used, the reinforcing bars are the same as in reinforced concrete. For the bed joints some special reinforcing elements like a plane truss were developed. Most of the structures are mixed constructions, using both masonry and reinforced concrete (see, Figure 19).

For multi-leaf walls and butt-jointed walls, wall ties are used. The wall ties are either regulated by the German code for masonry [1] or one needs a special certification. For example, see Figure 15.

3.2 *Thermal insulation*

The heating season, with outside temperatures at times below freezing, is more than six months in Germany. The potential of energy savings from thermal insulation is well established and has been regulated by code for a long time.

In Germany, a new heat insulation code exists effective since February 2002. It states that a maximum overall heat transfer value $U_{max} = 0,35 \, W/m^2K$ must be maintained in new buildings. This is only possible if masonry units with high thermal resistances and an appropriate wall thickness are used or a thermal insulation layer is added.

More voids or vertical cavities inside the bricks ensure that the higher thermal resistance is achieved.

The thermal insulation component usually consists of polystyrene foam panels or mineral fibre panels. These insulation layers are applied with varying thickness in the ceiling and the roof planes. Insulation is only effective if all exterior building components have been properly covered. Thermal bridges lead to energy losses and can also be the reason for damages to the building elements.

3.3 *Damp proof courses*

In Germany it must be verified that, that no water gathers within the walls throughout the year. This means that specific standards must be followed during the construction of the wall. Therefore, the vapour permeability, the temperature and absolute humidity of the materials, as well as their placement within the construction, play a decisive role.

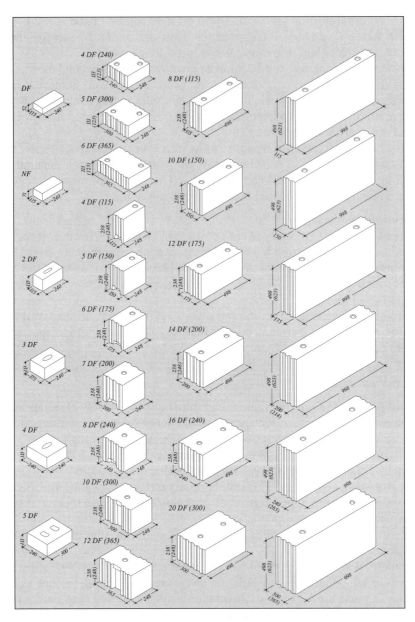

Figure 11.   Examples for calcium-silicate units [4].

Clinkers for façades have an especially high vapour migration resistance. When disregarded, this can lead to significant moisture damages. Based on this principle, the walls in newer buildings in Germany are generally ventilated through the use of an air space and weep holes, so that the water vapour can escape.

Masonry, when in direct contact with the ground, can be negatively impacted by moisture. There an impermeable condition is achieved through the application of

Figure 12.   Example of waterproofing for masonry [4].

a bitumen layer. Additionally, bitumen layers are inserted into the bottom courses for protection against moisture from capillary action.

3.4   *Wall finishes*

One possibility for a surface finish is the use of fair-faced masonry. In this case clinker walls act as the outer layer. The frequency with which fair-faced masonry is used as a façade element varies regionally. In Northern Germany, the so-called "brick houses" are the typical vernacular.

A less expensive alternative to the fair-faced masonry is the reproduction of bricks by queen closers. Similar to tiles, these are adhered on the brown coat.

The traditional finishing coats of lime or lime cement mortar are rarely used on outer walls today. Here pre-mixed mineral coats or synthetic resin coats with pigments are used. These are a part of the exterior finishing insulation system, which is applied in multiple layers and sometimes includes a layer of reinforcement and a layer of insulation. The finished façade then still gets a layer of paint applied.

Other kinds of architectural finishing coats are the raked finish, troweled finish or smooth finish which can be used for a particular aesthetic effect of the façade.

Further possibilities for the surface design are ceramic veneers. However, this is seldomly used because the vapour impermeability usually leads to building construction problems. Wood siding is also commonly used for façade treatments.

3.5   *Most common wall enclosures*

A multiple-layer wall is a common construction type for masonry. As illustrated in Figure 14, it consists of a load-bearing masonry layer, thermal insulation and finishing coats both on the interior and exterior.

Figure 13.   Building with a clinker façade [5].

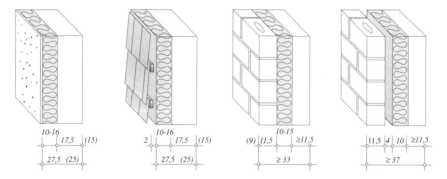

Figure 14.   Wall construction of a thermally insulated wall (units with additional insulation).

If the U-value of the available masonry units can be improved it will even be possible to eliminate the insulation layer as a part of the wall construction.

Historic buildings of sandstone are very common in Saxony, particularly in Dresden area. In these structures the entire wall consists of large sandstone blocks. Today the use of sandstone is very cost-intensive and thus new buildings are usually only applied with a sandstone veneer or a second non-load-bearing wall with fair-faced masonry is added.

Generally, reinforced concrete walls are used for the primary and load-bearing structure. For less prestigious buildings the use of sandstone is not considered a feasible option.

Figure 15.   Example of a multiple-layer exterior wall [4].

3.6   *Current problems*

For buildings under historic preservation it is not possible to retro-fit the structure with a thermal insulation system since the original façade generally needs to be preserved. Nevertheless, thermal insulation can be applied on the inside for energy-savings, though this can lead to moisture problems.

4   TRENDS AND DEVELOPMENTS

4.1   *Resolution for existing problems*

The dimensional stability of the masonry units used has been greatly improved within the last several years. Further, masonry unit-mortar-systems and mechanical aids have been developed which ensure a fast, simple and qualitatively better construction of the walls.

Therefore the safety factors for the structural calculations can be reduced since the quality control during production of the masonry units and during the construction is so consistent. Thus the masonry can be utilized more effectively and more economically.

Masonry units with large dimensions are heavy and usually difficult for the mason to work with. Ergonomic aids such as grip openings or grip holes have therefore been developed. These are placed in such a way that they are not visible in the finished masonry.

For economic reasons the masonry unit dimensions tend to continually be increasing. Beyond a certain dimension the masonry units can no longer be manually worked with. For this small cranes have been developed which can be moved quickly and easily.

Figure 16.   Masonry unit with grip openings [4].

Figure 17.   Example of building with large-sized masonry units.

The development of masonry units with greater percentages of vertical cavities and only thin webs, makes the application of the mortar for the bed joints more difficult. Mortar sleds were developed to simplify and accelerate this process. The sled is filled with mortar and pulled across the wall. Thus a uniform bed joint with a specific thickness is created. This is particularly important for masonry with a thin layer mortar.

Reinforced masonry is not commonly used in Germany. Its use is usually limited in combination with reinforced concrete, in which case specialized masonry elements are used as shells. Ring beams and lintels are most frequently used for this. Door and window lintels can be prefabricated elements.

Figure 18.   Use of mortar sleds for thin layer mortar.

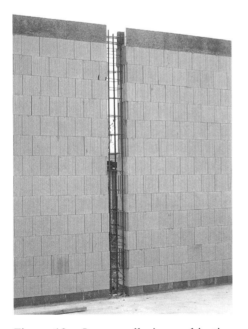

Figure 19.   Large walls in combination with reinforced concrete.

Figure 20.   Insulated U-elements for a ring beam.

Prefabricated elements are advantageous in concrete construction because of their short erection time. This is also the case with pre-fabricated masonry walls. As an additional benefit, no formwork is needed in masonry construction. Therefore, the production in small quantities is also economical.

Figure 21.   U-shaped clay masonry elements for a ring beam.

Figure 22.   Calcium-silicate pre-fabricated lintels.

Figure 23.   Pre-fabricated clay masonry lintels.

4.2   *Trends regarded to labour, quality, durability, productivity, etc.*

Two important goals in the further development of masonry construction are: the increase of the efficiency and improvement of the thermo-technical properties. It is also the aim to reduce the required labour and thus the costs for the construction of a wall, so as to make masonry constructions competitive with concrete, steel, and glass constructions.

Figure 24.   Prefabricated wall elements [5].

One possible approach for reducing labour costs is the development of dry masonry. This would eliminate man-hours during construction since joints would no longer need to be constructed. Furthermore, the material costs are reduced by the elimination of the mortar. A further advantage is the potential for being able to reuse the masonry units.

Yet other advantageous characteristics of masonry, such as its durability and the good Summer thermal heat insulation, should be further improved through research.

5   SIGNIFICANT REFERENCES AND DOCUMENTATION

[1] DIN 1053-1: 11.96: *Mauerwerk*. Teil 1: Berechnung und Ausführung. NABau im DIN, Berlin, November, 1996.

[2] DGfM – German Association for Masonry: http://www.dgfm.de/, 2002.

[3] Jäger, W.; *Typical Wall Structures in Germany*, Lecture for the CIB W023 Commission Meeting 2001 (unpublished manuscript), Dresden, 2001.

[4] Product Portfolio of the Calcium-Silicate Industry, Kalksandstein-Information GmbH + Co KG: http://www.kalksandstein.de/, 2002.

[5] Product Portfolio of the Company Wienerberger (Clay Units), Wienerberger Ziegelindustrie GmbH: http://www.wienerberger.de/, 2002.

[6] Schubert, P.: Schadenfreies Konstruieren mit Mauerwerk, Teil 2: Zweischalige Außenwände. In: Mauerwerk-Kalender 28 (2003). Ernst & Sohn, Berlin, S. 259–274.

[7] Pohl, R; Schneider, K.-J.; Wormuth, R.; Ohler, A.; Schubert, P.: *Mauerwerksbau – Baustoffe, Konstruktion, Berechnung, Ausführung*, Werner-Verlag, 4. Auflage: Düsseldorf, 1992.

[8] Belz, W.; Gösele, K.; Hoffmann, W.; Jenisch, R.; Pohl, R.; Reichert, H. – "Mauerwerk Atlas" 3. Auflage, München, 1991.

CHAPTER 6

Typical masonry infills for buildings in Greece

Elizabeth VINTZILEOU

Associate Professor
Faculty of Civil Engineering
National Technical University of Athens
Greece

SUMMARY

This paper summarizes information regarding the infill systems used in Greece in the vast majority of buildings. Data are presented regarding the importance of the building sector in Greece, the typical structural systems applied in buildings, the materials used and construction types applied for enclosure and partition walls. The main problems related to the pathology of infills due to earthquakes are mentioned. Finally, the perspectives of the brick production sector, as well as masonry infill construction are presented.

1 INTRODUCTION

Although the construction of structural masonry in Greece is very limited in the last decades, brick masonry is almost exclusively used for construction of enclosure and partition walls in buildings. In fact, brick masonry infills offer economic and durable solutions; they may ensure adequate insulation properties, whereas they contribute to the seismic behaviour of the structural system.

This last aspect is of major importance, since half of the total seismic energy of Europe is released every year in Greece. Theory and practice have repeatedly proved that masonry infills may affect in a positive way the seismic behaviour of buildings, provided that possible negative structure-infill interaction is avoided.

The aim of this paper is (a) to give an overview of the structural systems applied in Greece in the last 50 years, (b) to present the most commonly used materials and construction types for masonry infills, (c) to provide data regarding the behaviour of infills in buildings, as well as regarding the indirect infill design measures included in the current Greek Aseismic Code and (d) to give some information regarding the trends in evolution of the materials used and construction types applied for infill construction in the country.

2   THE CONSTRUCTION SECTOR IN GREECE

2.1   *The importance of the construction sector*

The construction industry constitutes the most important sector of economic activity in Greece. According to the data of the Institute for the Economy of Structures (IOK [1]), the total construction activity in Greece (public and private sector) for the year 2000 represents an amount of 10.3 billion euro. An increase of approximately 17.4% is estimated for the year 2001. 57% of the construction activity is directed to public works.

The contribution of the construction industry to the GNP is also increasing: It represented the 6.3% of the GNP in 1997, whereas this percentage was equal to 8.6% for the year 2000. It is estimated that for the period 2002–2004, the contribution of the construction industry will be as high as 12% of the GNP.

This development is attributed both to buildings and infrastructure (financed partly by the EU, as well as by the Greek State for the 2004 Olympics), as well as to the penetration of large construction companies to the private construction sector.

An index of the increasing productive dynamism of the construction sector in Greece is the following (IOK [1]): The average yearly increase of the GNP for the last 15 years is varying between 5% and 8%, whereas the respective increase of the total turnover of the construction industry varies between 10% and 22%.

This year, almost 350,000 people (27,000 of them, Engineers) are employed in the construction sector. The respective number for the year 1991 was 245,000 persons.

Based on data by the National Statistical Service of Greece (referenced by IOK), the total volume of private building construction was approximately equal to 70 million m^3 for the year 1990. It dropped to less than 50 million m^3 in 1995. Since then, it is continuously increasing and it is estimated that it will be over 80 million m^3 for 2003.

This figure includes both new buildings and additions (in-plan and/or in-elevation) to existing ones. According to a rough estimate, approximately 60 million m^2 of infills will be needed for enclosures and partitions. Therefore, the importance of infill walls becomes more than obvious, both from the economical and public safety point of view.

2.2   *Typical structural systems*

The structural system in the vast majority of buildings constructed after the Second World War consists of RC space frames.

As mentioned before, structural masonry was practically abandoned for various reasons, namely, the high cost of stone masonry which used to be the traditional material for structural masonry, the production of bricks with properties inadequate for structural masonry, the need for medium- and high-rise buildings in urban areas (which excludes structural masonry due to the seismicity of the country), the extensive damages suffered by old masonry structures during earthquakes, as well as, the lack (over a long period) of credible knowledge and Codes/Recommendations to regulate seismic design of structural masonry.

In addition, (a) the high cost of steel (mainly imported), and (b) on the contrary, the low cost of cement (produced in Greece) and the good quality of lime aggregates available all over the country, made reinforced concrete the "national building material".

As for the typical characteristics of the bearing system in RC buildings, one may distinguish three main periods (roughly '60s, '70s and '90s), taking into account both the geometry of buildings and arrangement of bearing elements, as well as the Codes applied for the design of structures. More specifically:

(a) In the '60s, the bearing system normally consisted of space frames up to 5 storeys high, with spans of the order of 3 to 4 m. Design was based on the 1954 Code for RC structures, as well as to the 1959 Aseismic Code. According to those Codes (based on permissible stresses concept), a very low conventional seismic force (between 4% and 12% of the total mass of the building) was applied to the structure, whereas measures to ensure sufficient ductility and to avoid brittle shear failures (common in the modern Aseismic Codes) were not prescribed. A rather dense arrangement of brick masonry infills was provided in all storeys.

(b) In the '70s, although the Codes were not modified, substantial changes were observed in the conceptual design of buildings. The high cost of the land, as well as the need for housing an increasing number of inhabitants in the urban areas led to an increase of the number of storeys (typically to 7, sometimes more). In order to meet the requirements of inhabitants for larger open spaces within their residence, the spans of the frames became larger (6–7 m). In addition, the need for parking places led to open ground floors (without any or with a limited number and length of infill walls). Moreover, the arrangement of infill walls in the remaining storeys became rather unsymmetrical, since larger lounges (with large openings) were arranged in the front part of the building, whereas bedrooms, bathrooms and kitchen were concentrated in the rear part of the building. Typically, shear walls were arranged to form the elevator shaft. They were, however, inadequately designed and detailed.

(c) '80s in Greece are characterized by a radical change in the Code for Design of RC structures, as well as in the Aseismic Code. The 1978 (Thessaloniki), the 1981 (Alkyonides) and the 1986 (Kalamata) earthquakes that affected large urban zones (namely, Thessaloniki, Athens, Corinth, Kalamata, etc.) led to the initiation of the process of elaborating new modern Codes (based on ultimate limit states concept). This process that passed by the parallel use (for some period) of the old and the new Codes was completed at the onset of the '90s. Therefore, the buildings constructed during the last 15 years respect more strict rules regarding the conceptual design of the structural system, they are designed for substantially higher seismic actions, they follow rules ensuring ductility (capacity design, confinement of concrete, adequate detailing, etc.). In addition, RC shear walls, adequately designed and detailed are provided in sufficient number to sustain a large part of the seismic action.

As observed, in the description of the structural system, infill walls are also mentioned, although they are usually considered as non-structural elements. However, when the design of a structure is governed by the seismic action (and this is the case in Greece), masonry infills do consist structural elements, since they contribute to the overall seismic behaviour of the structure. A short reference to the widely recognized effect of infills is given in the following section.

3   THE EFFECT OF INFILLS ON THE SEISMIC BEHAVIOUR OF RC STRUCTURES

Figure 1 summarizes (in a schematic way) the effect of infill walls on the seismic behaviour of RC frames.

As observed in numerous full scale tests (CEB [2]), the addition of masonry infills leads to substantial enhancement of the lateral resistance of the RC frame (by 50% to 500%, depending on the mechanical properties of the infill, on the relative frame to infill resistance, etc.), as well as to substantial enhancement of stiffness (by 100% to 3000%) which reduces the displacements imposed to the structure during the earthquake.

It was also observed that, even after extensive damage of the infills, at large imposed displacements, the lateral resistance of the infilled structure remains higher (by 0 to 100%) than that of the bare structure. It is also important to note that, thanks to the confinement provided by the surrounding frame, the angular distortion of the infill at maximum resistance increases to values as high as 0.005 to 0.008.

Nevertheless, the effect of infills is beneficial provided that the negative global and local interaction with the bearing system is avoided. In fact, as shown in Figure 2a, the non-symmetric in-plan arrangement of masonry infills within a practically fully regular bearing system leads to torsional effects that may cause premature failure of some bearing elements. Similarly, see Figure 2b, in case of a storey free of infill walls, the discontinuity of overstrengths observed along the height of the building results to concentrated ductility demands in the open storey. However, even in case

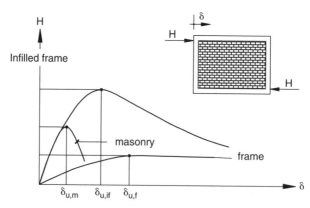

* enhanced resistance
* enhanced stiffness
* substantial residual resistance
* reduced imposed displacements
* increased deformation of masonry at failure

Figure 1.   The effect of infills on the lateral resistance vs. lateral displacement curve (schematic).

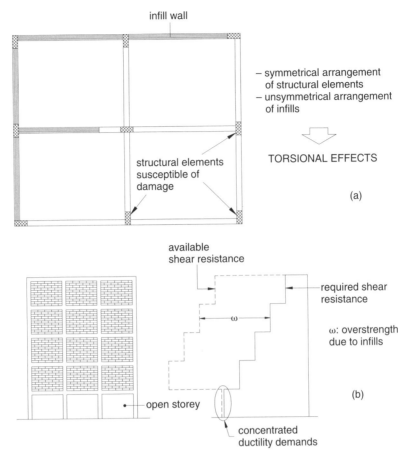

infill wall

– symmetrical arrangement
of structural elements
– unsymmetrical arrangement
of infills

TORSIONAL EFFECTS

structural elements
susceptible of
damage

(a)

available
shear resistance

required shear
resistance

ω

ω: overstrength
due to infills

open storey

(b)

concentrated
ductility demands

Figure 2.   The effect of unsymmetrical in-plane and in-elevation arrangement of infill walls.

of proper design and detailing, the columns of the soft storey may not be able to provide the required ductility.

The local interaction between frame and infill (Figure 3) should also be avoided. In fact, as shown in Figure 3a, due to the fact that the stiff infill tends to deform much less than the flexible frame, separation occurs between them at early loading stages, except in the region of the loaded corners of the frame. Thus, a diagonal strut is formed within the infill panel.

This strut imposes to the columns an additional shear force within their critical regions. In case those regions are inadequately detailed, premature shear failure may occur. Last but not least, in the quite common case of spans partly filled with masonry-to allow for daylight mainly in school and industrial buildings-the infill provides lateral support to the columns, thus making them to behave as (extremely shear sensitive) short columns (Figures 3b and 4).

The structural role of infills is recognized by modern Aseismic Codes, including the current Greek Aseismic Code (EAK [3]). Actually, (a) it is suggested to avoid

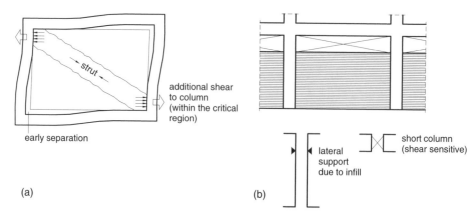

Figure 3.   Local effects of infill-frame interaction: (a) Additional shear force on the columns, (b) Formation of (shear sensitive) short columns.

Figure 4.   Typical short column failure due to: (a) discontinuous masonry wall and (b) inadequate reinforcement in the column (1999, Parnitha earthquake).

irregularities due to infills; (b) in case of an open storey, it is compulsory to arrange sufficient RC shear walls to restore stiffness and strength that would be provided by infills; (c) it is not allowed to modify the arrangement of infills in a building without the permission of a Structural Engineer (based on adequate checks), etc.

Specific measures, aiming at reducing damages of the infills themselves are also included. In this respect, it is required that the interstorey drift in any storey should not exceed 0.005.

Thus, infills (confined by the surrounding frame elements) are protected against extensive shear cracking. This check regards the serviceability limit state and, therefore, it is performed for an implicit design earthquake of shorter return period than for the structural elements of the building.

4   TYPICAL MASONRY INFILL CONSTRUCTION

4.1   *Introduction*

As mentioned before, the majority of infills in Greece are made of clay bricks. Nevertheless, there is also a small percentage of cases in which lightweight concrete blocks are used (with thin layer mortars).

In addition, in several office buildings, the external masonry infills are replaced by glazing, whereas light partitions substitute fully or partially the masonry partitions. In what follows, the typical case of clay masonry infills will be presented in detail.

4.2   *Common construction types*

Partition walls are typically 100 mm to 120 mm thick (depending on the dimensions of the bricks, Figure 5a) and they are plastered both sides. Partitions are in contact with the RC elements along their height. This contact is achieved by filling with mortar the gap between partition wall and columns.

Along their top, the walls are wedged to the upper beam or slab, by means of a layer of bricks placed inclined (Figure 5b). This last layer is placed after the wall settles (due to the shrinkage of mortar). In the most earthquake prone regions of the country, the tradition is to construct, at mid-height of the wall a RC tie beam (Figure 5b). The tie beam (height $\sim$ 100 mm, thickness equal to the thickness of the wall) is reinforced with 4 steel bars (often anchored into the RC columns). The tie beam is provided to improve mainly the out-of-plane behaviour of slender masonry walls during an earthquake.

As far as enclosure walls are concerned, one may distinguish two main cases.

In the buildings constructed before mid-seventies, the external walls were typically 200 mm to 250 mm thick and they were made by continuous bonding of clay bricks within their thickness (Figure 6a). They were plastered both sides. Their connection to the frame elements was as for partition walls.

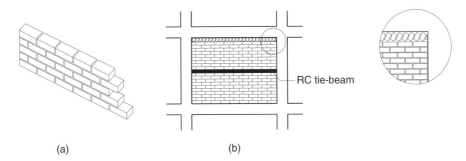

(a)   (b)

Figure 5.   Typical construction of partitions: (a) construction type; (b) attachment of the infill to the beam and the RC tie beam at mid-height of the infill.

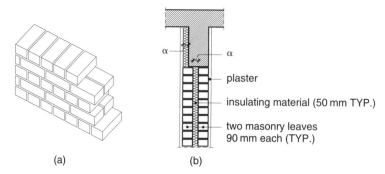

(a)   (b)

Figure 6.   Typical construction of external wall, (a) before '70s, (b) after '70s.

After the mid-seventies approximately, the construction type of external walls has changed. In fact, as shown in Figure 6b, within the same thickness as previously, a cavity wall is constructed. In the space between the two leaves, the insulating material is accommodated, with the exception of regions close to openings. There, the cavity is used to accommodate sliding doors and windows; thus, the effectiveness of insulation is reduced. The two masonry leaves are usually not connected to each other, unless a RC tie beam is provided.

4.3   *Materials for infill walls*

In Figure 7 some clay bricks (with horizontal holes) typical for infill construction are shown. It should be noted that bricks are produced by a large number of MSE all over the country. Thus, a quite large variety of bricks are available on the market. Some of them are used only locally, others are available in the whole country.

Depending on the geometry of the brick, as well as on the quality of clay, the compressive strength of the bricks varies between 1.5 MPa and 4 MPa (perpendicular to the holes).

The mortar used for construction is normally a lime cement mortar. Since masonry infills are not engineered, the mason does not need to follow certain mixing proportions for the mortar or respect some rules regarding the thickness of mortar joints. Nevertheless, in practice the following mix proportions apply usually: lime/cement/sand = 1/0.25/4 (by volume).

As mentioned before, usually, both enclosures and partition walls are plastered both sides. Plasters are applied in three layers, as follows (see also Figure 8): a) a layer with mix proportions identical to the mortar used for construction; b) a layer with mix proportions as follows: lime/cement/sand = 1.25/0.25/4 (by volume); c) a final layer on which paint is subsequently applied.

One of the following alternative compositions is applied: (1) lime/cement/sand = 1/0.5/4 (by volume) or (2) lime/marble powder = 1/2 (by volume).

The long experience related to the traditional plasters described here above has proved their satisfactory behaviour against moisture, weathering and fire.

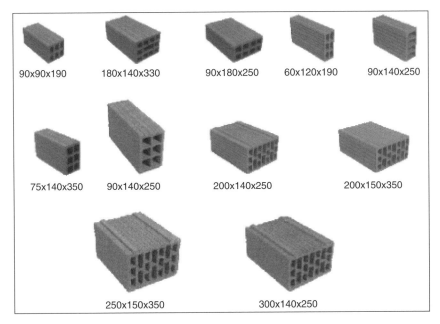

Figure 7. Typical clay bricks used for infills [dimensions in mm] (website of the Association of Greek Heavy Clay Manufacturers [4]).

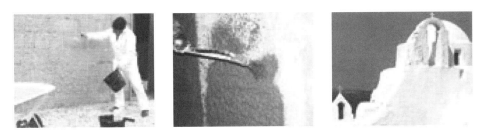

Figure 8. Plaster applied in three layers (website of Dionysos Marbles).

5 PATHOLOGY OF INFILL WALLS

Greece being an earthquake prone area, the most important aspect of pathology related to infill walls regards their seismic behaviour. In addition to the effect of infills on the seismic behaviour of the structural elements (described in Section 3), there is an issue of major importance both for public safety and economy.

In fact, failure of infill walls during an earthquake may injure or even kill people. Furthermore, the serviceability and damageability requirements may not be satisfied due to the extremely high cost associated with the damage of non-structural elements (infill walls, plaster, suspended ceilings, windows, doors, electric and sanitary fittings).

Tiedemann ([6] and [7]) who estimated the cost of several earthquakes related to damages in infill walls, concluded that the contribution of the non-structural elements to the total cost may be as high as 80%. This figure seems to be confirmed by Greek data. In fact, according to the information provided by the Organization for School Buildings (OSK [8]), 60% of the repair cost of school buildings damaged by the Parnitha earthquake (Sept. 1999) was related to damages suffered by infills.

Damages in infills are caused both by in-plane and out-of-plane seismic actions: For small imposed displacements, separation between infill and frames occurs (Figure 9). For larger imposed displacements, exceeding the one corresponding to

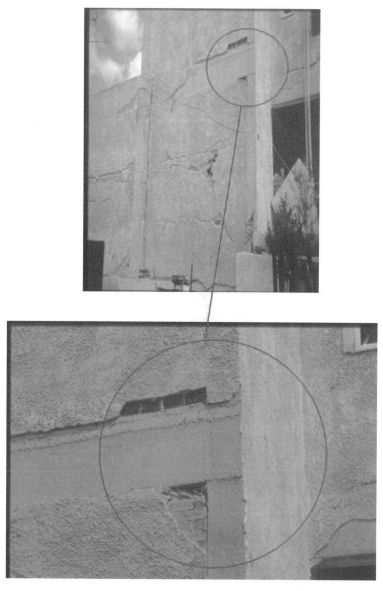

Figure 9.   Separation between infills and frame elements. Oblique bricks used to wedge the infill to the beam can also be distinguished (Parnitha earthquake, 1999).

the maximum resistance of masonry, shear cracks appear in the infill (Figure10). Disintegration and collapse of the infill may also occur (Figure 11).

It should be noted that the external cavity infill walls have proved to be extremely vulnerable to earthquakes. As shown in Figure 12, out of plane bending

Figure 10.   Shear crack in the exterior leaf of the cavity wall and out-of-plane collapse of external leaf (Parnitha earthquake, 1999).

Figure 11.   Shear cracking and collapse of the external leaf of cavity infill wall (Parnitha earthquake, 1999).

Figure 12.   Local disintegration and out-of-plane partial collapse of double leaf external wall (Parnitha earthquake, 1999).

Figure 13.   Out-of-plane collapse of the external leaf of cavity infill wall (Parnitha earthquake, 1999).

of the cavity infill (made of two thin low strength leaves) led to disintegration and partial collapse of the wall.

Figures 10, 11 and 13 show the typical damage observed in exterior cavity walls: Due to the lack of connection between the two leaves, the external leaf collapses. The insulating material placed in the cavity as well as the interior leaf of the infill can be seen on the pictures.

It has to be mentioned that in buildings constructed before mid-seventies, in which the exterior walls are of substantially smaller dimensions (therefore, less vulnerable to out-of-plane actions) and solid (it is reminded that they were constructed as shown in Figure 6a), very limited damages were observed.

Furthermore, in those cases, masonry infills played a very positive role in sustaining a large part of the seismic actions. In fact, as the structural elements were very flexible, as well as poorly designed and detailed, their contribution to the seismic behaviour of the buildings was rather limited.

6   TRENDS OF EVOLUTION

6.1   *Introduction*

To improve the double-fold function of infill walls, namely to ensure habitability and to improve the seismic response of the building and of the infills themselves, the following developments are under discussion in Greece, whereas some applications are already scheduled.

6.2   *Single-leaf enclosures*

As mentioned in the previous section, the external cavity walls are extremely vulnerable to seismic actions. In addition, the organic insulating material placed in the cavity is subject to ageing. On the other hand, the increasing sensibility of the society towards ecological issues leads to the re-invention of natural, inorganic materials for use in buildings.

In this respect, the use of masonry ensuring thermal, hygric and acoustic insulation within its thickness would be very advantageous. In fact, the Association of Heavy Clay Manufacturers in Greece is promoting the production and the use of bricks (see, for example, those of the last row in Figure 7) that are adequate for single leaf enclosures.

The resulting infill walls (approximately 300 mm thick, plaster included) satisfy the respective (national and European) regulations for insulation. In addition, the seismic behaviour of infills will also be improved. However, to improve the behaviour under earthquakes, additional measures may be needed.

6.3   *Use of reinforcement in infill walls*

Full scale tests carried out in the last two decades have proved that the addition of light reinforcement to infill walls leads to substantial improvement of their seismic behaviour both in- and out-of-plane. The reinforcement can be placed either in the bed joints of the infill (Figure 14a) or in both wall faces in the form of wire mesh (Figure 14b).

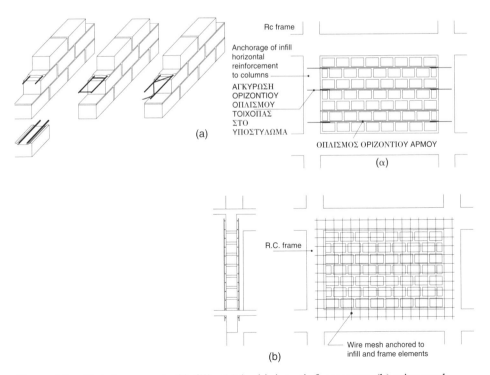

Figure 14.   Reinforcement of infills: (a) bed joint reinforcement, (b) wire mesh.

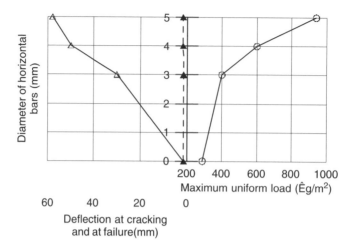

Figure 15.   The effect of bed joint reinforcement on the out-of-plane flexural behaviour of infills (based on experimental results by Vanmechelen et al. [12]). Masonry walls 2.7 m long, 1.5 m high, loaded by horizontal monotonically increasing uniformly distributed load.

As far as in-plane behaviour of infills is concerned, tests have proved ([9], [10]) that:

(a) Even a limited reinforcement ratio in bed joints improves the behaviour of infills at the stage before the attainment of their maximum shear resistance: The appearance of cracks is delayed and, thus, disintegration of masonry is to some extent avoided.
(b) In case double wire mesh is provided to the infill, the improvement is more spectacular over the whole range of imposed displacements.

Regarding the out-of-plane behaviour of infills, tests have proved ([11], [12], (see also Figure 15) that light bed joint reinforcement improves substantially the behaviour of infills at failure, as it enhances its bearing capacity and increases the deflection at ultimate.

On the contrary, it seems that cracking load and deflection at cracking are not affected by the presence of bed joint reinforcement. Here again, the addition of wire mesh on both faces of the infill leads to substantial improvement of the seismic behaviour for the whole range of imposed displacements.

It seems, therefore, that the addition of a small percentage of reinforcement in infill walls may lead to substantial gain both from the safety and economy point of view (in the sense of reducing damages due to earthquakes), without substantial increase of the cost of buildings. This is an issue widely discussed in Greece among Engineers. It is expected that in the next few years, the use of lightly reinforced infills will be rather common practice in the country.

7 REFERENCES

[1] Institute for the Economy of Structures (IOK), website: **www.iok.gr**, 2003.

[2] CEB – "RC frames under earthquake loading". State of the Art Report, Thomas Telford, London, 1996, 303pp.

[3] "Earthquake Planning and Protection Organisation (EPPO)". Greek Aseismic Code, NEAK", 2000, 257pp. (in Greek).

[4] Association of Heavy Clay Manufacturers (SEVK), website: **www.sevk.gr**, 2003.

[5] "Dionysos Marbles", website: **www.marmoline.gr**, 2003.

[6] Tiedemann, H. – "A statistical evaluation of the importance of non-structural damage to buildings". Proceedings, 7th WCEE, Istanbul, 1980, Vol. 6, pp. 617–624.

[7] Tiedemann, H. – "Structural and non-structural damage related to building quality". Proceedings of the 7th ECEE, 1982, Athens, Vol. 3, pp. 27–34.

[8] Organization for School Buildings (OSK), unpublished data, 2002.

[9] Zarnic, R.; Tomazevic, M. – "Study of the behaviour of masonry infilled reinforced concrete frames subjected to seismic loading – Part Two". Research Report ZRMK/IKPI-85/02, Institute for Testing and Research in Materials and Structures, Ljubljana, 1985, 234pp.

[10] Jurina, L. – "Pareti in muratura soggette ad azioni sismiche". Costruire, N.100, 1977, 46pp.

[11] Roussel, B. – "La maçonnerie armée. Cas des maçonneries de blocs creux en béton sollicitées en fléxion ». Thèse présentée à l'Ecole Nationale des Ponts et Chaussées, 1988, 121pp (plus Appendices).

[12] Vanmechelen, E.; Mortelmans, F. – "Reinforced Masonry Tests". Research Report, K.U. Leuven Research & Development, Belgium, 1988, pp. 75–90.

CHAPTER 7

Typical masonry wall enclosures in India

B. C. ROY

Executive Director
Consulting Engineering
Services (India) Ltd.
New Delhi
India

SUMMARY

According to the strategy defined on CIB Commission W023-Wall Structures and as
a contribution to the perspectives of masonry from across the world, an effort has been
made here to discuss masonry enclosure systems in India. The contribution highlights
the diversity in geography and culture and also the history of development of masonry
enclosures that have created the astounding variety. Recent developments with regard
to seismic performances are mentioned.

1 INTRODUCTION

In the context of an old, dynamic and a thriving civilization spread out over a vast
area, as in the case of India, it becomes essential to trace the evolution of a technology by considering the local conditions. India's every region offers such localized knowledge but with universal appeal.

If we want to discuss masonry enclosed systems in India we have to start at least
about four thousand years ago, the period of Harappan Civilization (Figure 1).
Brick masonry was used for foundations of houses, structures for drainage of rainwater and sewerage, baths with raised brick platforms, and stone masonry as flood
barriers. This is an unbroken tradition.

Beyond antiquity, the other important fact to be considered is India's truly sub-
continental geographical extent. The weather patterns range from places of heav-
iest rainfall worldwide to scorched desert environment to snow clad mountains to
a long shoreline to almost tropical forests. And people lived and live in all of these
places and their homesteads in many forms, which is what masonry structures typically enclose, exhibit this variety. For every major epoch in Indian history we
have a distinct style, a specific set of materials and so on.

Figure 1.   Harappan Civilization.

Figure 2.   Sun Temple, East India.

Figure 3.   Writer's Building Kolkata.

Figure 4.   Chhatrapati Sivaji Terminus Mumbai.

India's spiritual moorings have given rise to, through the ages, magnificent temple structures with large enclosed areas, many of them of stone masonry with decorative facades: the Shore temple in Mahabalipuram (8th–9th centuries), the Sun Temple in Konarak (13th century) (Figure 2), and all the temples dotting the banks of the great rivers like Yamuna, Ganges and Brahmaputra.

Coming to the more modern times, the pre and the colonial era, Taj Mahal stands tall amongst masonry enclosures, stone in this case. In slightly later times, in the year 1786 the Writer's Buildings in Kolkata (Figure 3) was built, followed by the Chhatrapati Sivaji Terminus in Mumbai (Figure 4) and of course, the buildings of Lutyen's Delhi, including the Rashtrapati Bhawan (the Presidential Residence) (Figure 5).

One has to appreciate the masonry artwork done in such loving detail that the buildings stand testament by their sheer longevity. Whether working with brick, concrete block, tiles, terra-cotta or stone, the skill and the quality concerns of the masons are much in evidence.

One can see masonry enclosures as small houses, or medium to large residences and other public buildings throughout the land.

The usually high ceiling of masonry enclosures is an environmental necessity, to keep the enclosures as cool as possible during the hot summers of the tropics and semi-tropics. Figure 6 shows a typical interior of this type of building. In the southwestern region that bears the brunt of the southwest monsoon, steep angled pitched

Figure 5.   President's Residence New Delhi.

Figure 6.   Chettinad, Chennai.

Figure 7.   Pitched   roof   house, Thiruvananthapuram.

Figure 8.   Temple Building Himalayan Region.

roofs (Figure 7) of tiles, either one or two storeyed, is the traditional form of housing. In the northeastern states, bamboo with mud walls is used extensively. In the reaches of Himalayas and also the Western Ghats (edges of the Deccan plateau) one sees the predominance of stone masonry that needs no explanation. Figure 8 shows a temple building in the Himalayan region.

Masonry construction bears loads, divides space, provides acoustic and thermal insulation, offers fire protection and also attractive facades. Its varied utility, therefore, is one of the main reasons for its survival through the years. Exteriors of a residential building wherein the masonry walls, of high strength and fine, textured finish, highlight the strength and beauty that can be achieved in this type of construction.

2   TYPES OF MASONRY ENCLOSURES

2.1   *Introduction*

While the diversity in India was mentioned with regard to geography and climate, the other important classification would be with regard to urban settings and rural life. The most visible difference between the two is that in the former one would witness multi-storeyed residential premises whereas in the rural they would typically be single-storeyed or at the most two storeys. Of course, there are communities that might defy a simple classification of this sort and the distinctions do dutifully fade away. In the following this divide is acknowledged.

Hold of tradition in urban areas is a little looser as can be understood. On an average, one might say that the older the form and the construction, the masonry enclosure is more likely to be in the rural areas. Add to this, about 2/3 of Indian population resides in rural areas. However, traditional forms and even materials find use even today in the urban areas, especially amongst the lower strata of society. Various types of housing stock are described below and the context, urban or rural, and the strata of society that uses a particular form have to be understood keeping the above in mind.

2.2   *Mud house with pitched roof*

This is a typical rural construction found throughout India, except in the high rainfall areas in the northeastern part of the country. It is a single-family house, mainly occupied by the poorer segment of the population.

The main load-bearing system consists of mud walls, which carry the roof load. In some cases wooden posts are provided at the corners and at intermediate locations. Structural integration between the wooden posts and the walls is lacking. There are only few openings (doors and windows) in these buildings. In general, this type of construction is built by the owners and local unskilled masons and the craftsmanship is very poor. This is a low-strength masonry construction and its attendant weaknesses are evident. Figure 9 shows a view of typical mud house.

2.3   *Timber frame brick house with attic*

This type of construction is suited mainly for residence and found throughout India, with variations to suit different cultures, material availability and environmental considerations. Timber is primarily used for the frame structural elements but due to the acute shortage of timber, and also perhaps of the awakening of ecological conscience, this construction type is in declining practice.

Timber frames, placed in longitudinal and traverse directions, are filled with brick masonry walls. The floor structure is made of timber planks. Most of the buildings are found to be rectangular in shape with few openings. The roofing material is usually galvanized iron sheet.

Figure 9.   Mud House.

Existing old structures, however, require maintenance and strengthening. The cost of this type of construction will increase if features relevant to the level of technological sophistication demanded by societal concerns, for example, safety against earthquakes, are to be incorporated. Of course, the returns for the increased outlay are the increased performance of the building.

It is found that comparatively older buildings are generally made with sufficient precision and good building skills. Extensions or modifications, however, are generally done without appropriate building construction techniques. In urban areas, the outermost rooms are converted into shops or used for commercial purposes. It has been found that some modifications and extensions have been done with heterogeneous materials and structural systems.

Views of typical building, timber frame and modification like continuation of an existing brick wall with random rubble masonry, etc. are shown in Figures 10, 11 and 12. The older houses are made up with mud mortar but recent extensions are built with cement mortar. Figure 13 shows a sketch of the structural elements.

Figure 10. Timber frame. House with masonry frame & iron sheets.

Figure 11. Building with timber frame.

Figure 12. Inappropriate and heterogeneous modifications.

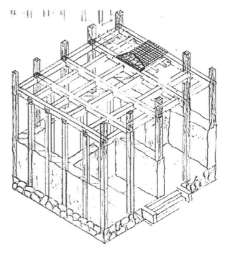

Figure 13. Sketch showing structural Elements.

2.4   *Rubble stone masonry walls with timber frame and timber roof*

This typical rural construction in central, southern, and northern India houses millions of people. It is cheap to construct, using fieldstones and boulders. The load-bearing structure is a traditional timber frame system, known as "khan". It is a complete frame with timber posts spanning about 2.6 m. Typically 600–1200 mm thick stonewalls provide enclosure and partial support to the roof.

Walls are generally supported by strip footings of uncoursed rubble masonry. The roof structure consists of timber planks and joists. To help keep the interiors cooler during hot summer (peak temperatures exceeding 40°C), a thick mud overlay covers the top the roof.

In general, the buildings of this type have undergone modifications over time. Figure 14 shows typical buildings and structural details including the timber roof. They are mainly built around the central courtyard and can be expanded horizontally by building additional rooms.

In some cases, there is a vertical extension. However, this is not very common. Also, in the aftermath of earthquakes in the '90s in the region where this type of construction is highly prevalent, there has been a general trend of removing heavy roofs in the buildings.

2.5   *Unreinforced brick masonry walls in mud mortar with flat timber roof*

This is a traditional construction practice prevalent both in the urban and rural areas of northern India, particularly in the western part of the Uttar Pradesh state in the Gangetic Plains. According to the 1991 Indian census, this construction constitutes

(a)   (b)

(c)

Figure 14.   Rubble stone masonry walls with timber frame.

about 17% of the total national housing stock and about 31% of the Uttar Pradesh housing stock.

Typically, this is a single-storey construction. The main load-bearing elements are unreinforced brick masonry walls in mud mortar. The roof structure consists of timber beams supported by walls.

Clay tiles or bricks are laid atop the beams; finally, mud overlay is placed on top of the tiles for thermal protection and also to prevent leakage. As such there are few modifications in this type of buildings. The only modifications that are carried out are in terms of providing extensions by constructing one room in the over the terrace of the housing unit. Figure 15 shows a typical building.

2.6 *Unreinforced brick masonry walls with pitched clay tile roof*

This is a centuries old Indian construction practice. Buildings of this type are used for residential, commercial, and public purposes throughout India, but especially in the northern and central parts, where good quality soil suitable for brick is widely available. This is a single-storey construction used both in rural and urban areas. The walls are constructed using clay bricks laid in mud, brick-lime or cement/sand mortar. The roof does not behave as a rigid diaphragm.

Figure 16 shows a typical building while Figure 17 shows key load bearing elements in roof construction.

Figure 15.   Unreinforced brick enclosure with flat timber roof.

Figure 16.   Unreinforced brick enclosure with pitched clay tile roof.

Figure 17.   Roof Elements.

Figure 18.   Unreinforced brick enclosure with reinforced concrete roof.

Figure 19.   Bhonga.

2.7   *Unreinforced brick masonry building with reinforced concrete roof slab*

This housing type is used for typical rural and also urban construction all over India. This construction is widely prevalent among the middle-class population in urban areas and has become popular in rural areas in the last 30 years. Brick masonry walls in cement mortar function as the main load-bearing elements. The roof structure is a cast-in-situ reinforced concrete slab. Figure 18 shows a view of a typical building.

In urban areas, additional floors are often added, sometimes even without considering structural aspects. The construction is therefore staggered and a gap of several years may exist between the construction of different portions of the building. In rural areas, where the habitats are generally spread out, horizontal building expansion is more common.

2.8   *Traditional rural house in Kutch region of India (Bhonga)*

The Bhonga is a traditional construction type in the Kutch district of the Gujarat state in western India. A bhonga consists of a prismatic enclosure with conical roof single cylindrically shaped room.The bhonga has a conical roof supported by cylindrical walls.

Bhonga construction has existed for several hundred years. This type of house is quite durable and appropriate for prevalent desert conditions. Due to its robustness against natural hazards as well as its pleasant aesthetics, this housing is also known as "Architecture without Architects." Figure 19 shows typical buildings and Figure 20 shows its roof.

Recent Bhonga constructions have used wide variety of construction materials. These include stone or burnt brick masonry either in mud or cement mortar.

Traditional roof consists of light-weight conical roof, while some recent constructions have used heavy mangalore tiles on roofs and strip footing below the wall, while traditional construction simply extended the walls below ground level. This body-plan is scalable with modular units laid abetting the existing ones with appropriate exit and entry. A concept plan is shown on Figure 21.

2.9   *Brick masonry – residential and other buildings*

Depending on the strength of the bricks, up to three storey high residential buildings can be, except for the ceilings, true and exclusive masonry enclosures. The

Figure 20.   Bhonga – inside view.

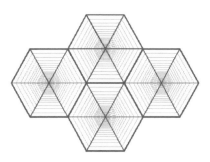

Figure 21.   Scaling of Bhongas.

Figure 22.   Residential building with exposed brick load bearing wall.

Figure 23.   Steel frame with masonry infill-residence.

requirements are high strength and high quality bricks and also stringent quality consciousness during execution. Architectural flourishes can be accommodated by special load bearing bricks.

This type of construction is usually found in the localities of the upper strata of society for the characteristics mentioned above (Figure 22).

2.10   *Reinforced concrete frame building with masonry infill*

The construction of reinforced concrete frames with brick masonry infill walls has been a very common practice in urban India for the last 25 years. They are finding increasing use in semi-urban (big towns) and even in some villages.

Typical modifications found in the case of a multi-storey building are in the form of alteration of position of interior walls. Additional storeys are added on many old one or two storey buildings without considering the load-carrying capacity or behavior under earthquake loading. Open balconies are also often enclosed in RCC buildings to increase size of rooms or to provide additional rooms. Figure 23 shows a typical building.

Many recent public buildings, like colleges and other institutions sport this construction, many times adorned with brick facia treatment (Figure 24).

Figure 24.   Reinforced concrete frame with masonry infill-business.

Figure 25.   Masonry infills with aluminium cladding.

Figure 26.   Low-strength dressed stone masonry building.

An extreme form is a multi-storey building with aluminium cladding and a glass façade (Figure 25). This has become a symbol for the corporate offices, especially of the new industries.

2.11   *Low-strength dressed stone masonry buildings*

Construction of stone masonry buildings using easily available local materials is a common practice in both urban and rural parts of India. Stone masonry houses are used by the middle class and lower middle class people in urban areas, and by all classes in rural areas. Tellingly, it finds quite extensive use in the city of Jaipur, a mid-sized city and the capital of the western state of Rajasthan.

In rural areas, these buildings are generally smaller in size and are used a single-story, single-family housing. In urban areas, they are up to 4 stories high and are used for multifamily housing. This is a typical load-bearing construction, in which both gravity and lateral loads are resisted by the walls supported by strip footing.

If the locally available stone is soft, dressed stones are commonly used and can be chiselled at low or moderate cost. Mud or lime mortar has been used in traditional constructions; however, recently, cement mortar is being increasingly used.

Because soft sandstone is readily available in the Kutch region of Gujarat, stone block masonry constructions are widely used for both single- and multi-storey constructions with a gable end timber roof truss or RCC roof slabs. Generally, local artisans without formal training are deployed in the construction and the results are structurally weak. Figure 26 shows a typical building.

3   MATERIALS

Masonry could have only meant rubble or stone masonry in the earliest times. It is impressive then that this form of homestead building has endured literally many thousands of years. What more, they offer stiff competition to the more recently developed systems like brick and hollow block masonry.

The most frequently used masonry products are clay bricks and concrete blocks. Brick masonry has evolved over many millennia and has also given rise to variations that have spurred innovations.

We may call bricks as the first reinforced material, as chopped straw and grass were mixed with clay to reduce distortions and cracking. Burnt, whether under the sun or in kilns, clay bricks, have withstood the test of time and is a staple of housing construction, either as load bearing or non-load bearing walls like partition and filler walls in India.

Choice of masonry units considers local availability, compressive strength, durability, cost and construction ease. Brick is favoured to stone because it is available at reasonable cost and of good quality. In hills and in some plains, stone is preferred as the soil is not suitable for making good quality bricks.

Concrete blocks may be economical in these regions if available stones require considerable effort and cost to be dressed to suitable shape and size. Sand-lime bricks – not common in India – are preferred in areas where sand and lime are available in abundance and any of the above features proves costly. Figure 27 shows a typical section of a brick wall adopted in India with normal bands.

Strength and size of bricks vary across the length and breadth of the country. There is a wide variation in the strength of the bricks, from 3 N/mm^2 to 20 N/mm^2, and also in the sizes – $230 \text{ mm} \times 115 \text{ mm} \times 75 \text{ mm}$, $240 \text{ mm} \times 120 \text{ mm} \times 70 \text{ mm}$, $230 \text{ mm} \times 100 \text{ mm} \times 45 \text{ mm}$, $225 \text{ mm} \times 112 \text{ mm} \times 56 \text{ mm}$, $210 \text{ mm} \times 100 \text{ mm} \times 70 \text{ mm}$, $220 \text{ mm} \times 110 \times 70 \text{ mm}$, $190 \text{ mm} \times 90 \text{ mm} \times 60 \text{ mm}$, $222 \text{ mm} \times 106 \text{ mm} \times 67 \text{ mm}$, and $229 \text{ mm} \times 108 \text{ mm} \times 72 \text{ mm}$ etc.

The tensile stress is limited between 0.10 N/mm^2 and 0.07 N/mm^2. Efforts are on to standardize the shape and size of the brick throughout the country. In certain parts of the country, especially in urban areas, machine made bricks having a compressive strength ranging from 17.5 to 25 N/mm^2 are being produced.

The normal progression beyond brick masonry is cavity walls. When originally developed, cavity walls consisted of two separate brick or stonewall with about a 50 mm air space between them. Cavity walls were developed to reduce the problems associated with water seepage (Figure 28). Water seepage cannot jump

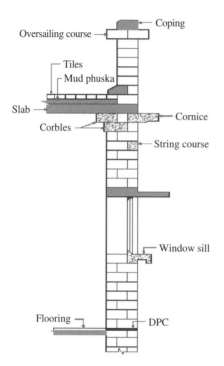

Figure 27. Normal bands as per Indian Standard codes in brick wall.

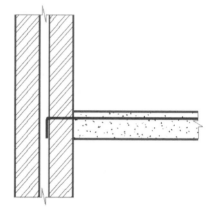

Figure 28. Cavity Walls and typical connection to RC Slab.

across the gap between the outer and inner walls, and the enclosure was thus water proofed.

Cavity walls could also support heavy loads such as a roof or floor. In 1850 a special block with air cells was developed. Over the years modifications to this product were introduced until the industry arrived at the level of standardization we see today.

With the subsequent advent of cement and concrete cavity walls of bricks morphed into hollow block panels. In many of the multi-storeyed buildings, hollow blocks are used as filler materials within the frame skeleton.

4   DESIGN PRACTICE

In India, the Indian standards prevail across the nation. However, the regional variations are accommodated in documents, compiled after extensive research and consultations, like the National Building Code, Central Public Works Department (CPWD) codes, State PWD codes. These guide the engineers in designing and building stable, safe and sustainable masonry enclosures.

Masonry codes in general covers structural design aspects of unreinforced walls that bear loads or not bear (like partition walls), constructed with solid or perforated bricks, stones, blocks, etc., permissible stresses and design methods. No provision has been stipulated for masonry in mud mortar.

In load-bearing brickwork, stability is gained from stipulations on the maximum-cantilevered distance. In brickwork not bearing loads these rules can be exceeded by reinforcing with steel fixings tied back to the structural framework. The load carrying capacity of the walls is also affected by its slenderness. To limit slenderness lateral supports are intended to reduce buckling and ensure stability against overturning.

Ensuring a good box action between all the elements of the building, that is between roof, walls and foundation renders seismic resistance to the building. For example, a horizontal band introduced at the lintel level ties the walls together. Further, walls also need to be tied to the roof and foundation to preserve their overall integrity. In a similar manner, engineers are advised to balance the mutual orthogonal strengths and weaknesses of intersecting walls.

A severe geometric discontinuity like a corner must have good interlocking. Openings (Figure 29) assume significance in deciding the performance of masonry buildings in earthquakes. Large openings also create impediments in the flow of in-pane forces and recommendations in the documents address such concerns.

Adequate gap is necessary between different blocks of the building. The Indian Standards suggest minimum seismic separations between blocks of buildings. However, it may not be necessary to provide such separations between blocks, if horizontal projections in buildings are small, say up to 15–20% of the length of building in that direction.

An arch is simply a way of supporting a wall above an opening and gives a pleasing look if properly constructed. The simplest arch is formed with uncut headers and the tapered joint filled with mortar. It was used in older construction but due to its susceptibility under earthquake disturbances, the practice has lessened in present masonry buildings and is being discouraged in the light of safety considerations.

Arches, besides been architectural embellishments, also served a functional purpose in the provision of high ceilings. However, in the current day context the volumes that buildings enclose are minimized and hence arches are out of favour.

Corbelling, that is use of brackets, projecting beyond the face of a wall acting as cantilevers are being adopted (Figure 30). In load-bearing brickwork, the maximum cantilevered distance of the corbel is determined by the one third rule, that is, each brick should not project more than one third of its bed length, and the maximum distance corbelled by a number of bricks should not exceed one third of the wall's width. In non load-bearing brickwork these rules can be exceeded by reinforcing with steel fixings tied back to the structural framework.

Not too many tall load bearing masonry buildings were constructed in the past due to variable size of bricks, and their quality. However, mechanized brick making has made it possible to construct 5–6 storey load bearing structures in some parts of the country, at a lesser cost as compared to RCC framed structures. With this development, structural design of load bearing masonry structures has assumed additional importance in India.

Five storey masonry block buildings in one brick thickness have been constructed in 1975 on an experimental basis by the National Building Organization. It may be noted that giving due deference to local practices the brick sizes were kept at 25.4 cm in Kolkata and 22.9 cm in Delhi.

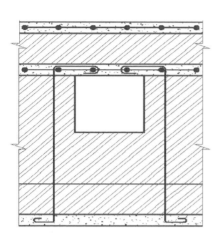

Figure 29.   Opening.

Figure 30.   Corbel.

Figure 31.   English Bond adopted
in India.

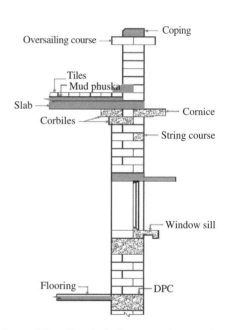

Figure 32.   Bands below openings and
at sill level from seismic consideration.

The bond adopted in general is the English Bond (Figure 31). Mortars used are cement-sand mortar with a maximum ratio of 1:3, lime mortar, cement-lime mortar and is adopted.

Figure 32 shows the bands below openings and at sill level from seismic consideration.

Figure 33 shows behaviour of a two storeyed building constructed using cement blocks, with continuous lintel band and corner RC columns.

The building developed extensive cracks below the lintel band while no cracks were developed above it. The corner columns suffered severe damage. Lintel band protected the portion above by creating a rigid block.

The damage below the lintel band indicate need of horizontal band at sill level. Provision of RC column at the corners or elsewhere, without bonding with the infill wall causes discontinuity in masonry walls leading to the building losing its integrity. Refer to Figure 34.

A school building with typical out-of-plane failure of walls between two openings is shown in the Figure 35.

Studies have shown that the strip of wall between two large openings experiences large deformations during flexural vibrations. A well dressed stone masonry building with cement mortar and light ferrocement roofing behaved well with very little cracking (Figure 36).

Figure 33.   A two-storey building.

Figure 34.   Wall separated from.

Figure 35.   Collapse of wall between openings.

Figure 36.   Well dressed stone in CM with ferrocement roof.

It was found that masonry buildings in mud mortar or lime mortar are prone to severe damage due to lack of bond strength essentially the result of out-of-plane flexure while the same with cement are better.

Use of lintel band, recommended by the Bureau of Indian Standards appears to introduce a rigid box-like behaviour in the upper portions of the building while the portion below the lintel band is cracked badly.

Additional horizontal bands, possibly at the sill level and at plinth level, are needed. These bands are helpful in tying the walls together at the junctions and also in preventing the growth of vertical cracks and in-plane shear cracks.

Provision of corner reinforcement in corners and junctions has to be properly-bonded with the surrounding masonry possibly with dowels or keys to prevent separation (Figure 35).

5   EVOLUTION OF MASONRY TO COUNTER SEISMIC EVENTS

It is only natural that a thousands year old technology undergo revolutions from time to time as well as be subjected to the slow process of evolution. It is only in recent times, say the past 200 years or so, science (marked by a rational analysis) and technology (supported by experience and empiricism) have come together in building sciences; before then it was mostly empirical as in "this has worked and hence OK" and "this has failed and hence not OK". One of the major issues of this fusing of science and technology is in earthquake effects on buildings and efforts to counter it.

It is in the above context we may be able to understand the evolution of the various structural and construction systems for masonry enclosures mentioned earlier. Traditional building systems sustain themselves even against deficiencies in a scientifically underdeveloped environment. People, even after devastations tend to resign themselves to, "we don't know what happened but let us build ourselves anew". The old systems, with their weaknesses persist. In India, this mentality may be more explanatory as to how structural deficiencies have not been weeded out.

It is generally agreed that newer technologies that render a structure a robustness that withstands nature's fury, say an earthquake, have to be inducted into the traditional systems of construction. Here a rural, urban dichotomy presents itself. The holds of traditions are a little loser in the urban areas and also the resistance to new ideas, little softer.

However, in the rural areas tradition holds a stronger sway. As engineers, we then have to find ways to infuse safety measures while at the same time not bring drastic changes in the style and functions of masonry enclosures. This, along with appropriate levels of community involvement, is the only way safety improvements can be accommodated. In this section, an effort will be made to locate the significant points of structural weaknesses vis-à-vis earthquake incidence and later, some examples. It might be beneficial to resort to case studies as and when appropriate.

The structural deficiencies are quite evident to the trained eyes of a structural engineer, but not necessarily so for the lay people. Most of the structural forms/ construction systems suffer from one weakness or the other.

For example, good workmanship alone, in the case of timber frame brick house, gives this structural form certain strength and rigidity. However, it is found that ad-hoc additions and extensions, some out of commercial considerations, subtract from the integrity of the original construction.

Heterogeneous materials and structural systems contribute to the weaknesses of the building. In general theses are modifications like continuation of an existing brick wall with random rubble masonry. It must be said here that cost would be increased only nominally should earthquake resistant features are introduced or resistance sapping practices are avoided.

Rubble stone masonry walls with timber frame and roof suffers from a more serious structural deficiency: heavy roofs and poor wall construction. Walls are either supported by strip footing or unfounded – lack of structural integration and a major source weakness in the event of a ground shake. It is to the credit of the people that after the 1993 earthquake in the state of Maharashtra, there has been a general trend of doing away with heavy roofs.

Reinforced concrete frame buildings with masonry infill walls are designed for gravity loads only, in violation of the stipulations of Indian Standards for earthquake-resistant design. These buildings performed very poorly during the Bhuj earthquake of January 2001 and several thousand buildings collapsed. The collapse was not limited to the epicentral region. The seismic vulnerability of this construction is clearly demonstrated by the collapse of a number of RCC frame buildings and damage to several thousand others in and around Ahmedabad, which is over 250 km from the epicenter. Figure 37 shows seismic strengthening details activity.

Reinforced concrete frame buildings with masonry infill walls, if constructed as per the stipulations of the appropriate codes of practice, are structurally strong under seismic conditions. But, commercial considerations play a major part in compromising on the safety with an it-won't-happen-to-me attitude. But, when it did on January 26, 2001, in the state of Gujarat, everyone suffers – loss of life and property of millions of rupees worth.

In the aftermath of this devastation there has been a concerted effort to stipulate more robust norms and also tighten the procedure for monitoring construction. The fact that these are after-the-fact efforts takes a little away from the shine off the structural engineering profession in India.

Unreinforced brick masonry wall, with flat timber or pitched clay tile roof, is a weak construction but finds a clientele amongst the rural and urban people all across India.

The main seismic deficiencies are, much like for the rubble stone masonry walls with timber frame, heavy roofs and weak walls. Lack of structural integration amongst the elements is an added negative feature. Fixing of a continuously running seismic belt (Figure 38), in the case of reinforced concrete roof, a continuous lintel, (Figures 39 and 40) is a strengthening measure that would pay for itself in terms of protected lives and property.

Again, driven mainly by economic considerations, additional floors are added ad-hoc without a proper evaluation of the structural aspects. Construction may be staggered in time and this in itself is a source structural weakness. Efforts are on to make knowledge of such subtle aspects widespread.

Figure 37.   Seismic strengthening.

Fixing of seismic belt

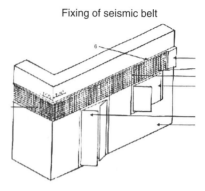

Figure 38.   Seismic strengthening – Fixing seismic *belt.*

Figure 39.   Seismic strengthening – Lintel bandage.

Figure 40.   Seismic strengthening – Corner Strengthening.

It cannot be readily concluded that lighter roofs offer better seismic resistance. Heavier roofs, like flat RC slabs offer a lateral integrity to the walls and this also enhances seismic response. The engineer, in such a case, is required to balance the competing claims and offer the most appropriate solution.

The rural mud house with pitched roof does not exhibit even the concept of structural integration and is extremely weak against ground shaking. But as these house are for the poor who are able to afford no more than the bare minimum, it would take an extraordinary effort to get them to listen to planning for the future, as in incorporating safety features. This gives rise to a vicious cycle. Efforts are on to create a cheaper/affordable and safe structure.

In this context, mention must be made of Bhongas, in the Kutch region of the state of Gujarat. Bhongas performed exceedingly well during the 2001 earthquake.

Very few bhongas experienced significant damage and the damage that did occur can be attributed to poor quality of construction materials or improper maintenance. It is still better that very few injuries were caused by the few bhongas collapsing.

The low strength dressed stone masonry buildings are usually built by local artisans without formal training and perhaps for this reason the completed structures are weak and are incapable of resisting seismic forces. Thousands of these houses collapsed during the 2001 earthquake and it may be concluded that these are inherently unsafe in areas of moderate to high seismic hazard.

6   AFTER AN EARTHQUAKE. A RECONNAISSANCE

6.1   *Introduction*

Jabalpur a city in central India experienced a moderate earthquake on May 22, 1997. The city and several villages located near the epicentre suffered heavy damages. In urban areas 40000 houses were partially damaged and 6000 rendered unsafe for living. The Department of Earthquake Engineering at University of Roorkee, India sent a team to survey the damage and the section below presents a reconnaissance.

The structures in the city comprised Un-Reinforced Masonry (URM) construction, mixed constructions employing Reinforced Concrete (RC) slabs and/or frames in one direction and load-bearing brick walls in the other, and other types of constructions.

A large number of URM structures existed near the epicentral region, and URM being extremely vulnerable to earthquake forces and as a result these structures suffered extensive damage. Load-bearing brick walls containing openings for doors and windows creating systems of wall piers and spandrels were observed to fail in shear, developing diagonal cracks. Also many out-of-plane failures arose from dynamic instability of slender walls. Slenderness ratio being an important criteria in all codes, as discussed earlier needs a proper adherence.

RC frame buildings exhibited superior performance in comparison with URM structures. However brick infill walls experienced severe cracking, and falling bricks posed a serious hazard. Many structures taller than two storeys were damaged but soundly constructed two storey houses of regular form fared better.

The majority of new buildings in the region had weak RC frames with solid brick infills and survived with minor damage. When properly constructed these buildings exhibited superior behavior as compared to non-engineered structures. Beam-column frames are designed for gravity loads only and are weak for lateral loads. The frames may not be present in both orthogonal directions. Often, RC portal frames support the floor over open bays in one direction with load bearing brick walls in the other. Failures in RC beams or columns were rare.

Unreinforced brick infills failed in in-plane shear and were knocked out-of-plane due to lack of anchorage to surrounding frames. These walls attract more in-plane seismic forces than the flexible and weaker RC frames and crack at relatively low seismic force levels.

6.2   *Structural deficiencies*

It is difficult to state a precise cause for most structural failures as often several deficiencies contribute simultaneously.

Errors in conceptual design were common, particularly caused by designers neglecting earthquake-resistant design practice as documented in Indian Standards. Plan irregularity exists in buildings and more damage occurred in buildings with rectangular plans. Lack of continuity at re-entrant corners cause large tensile stresses to be develop in the floors. Stair openings located in these areas also accentuate the problem.

It is difficult to remedy plan deficiencies satisfactorily after construction, but provision of horizontal struts ("drag" members) to strengthen load paths connecting vertical structural members can significantly improve seismic performance.

Diaphragms should be checked for tensile stresses near re-entrant corners and poorly placed staircases should either be relocated or isolated from the rest of the building. One 5-storey building was constructed with a weak RC frame and infill walls. Vertical discontinuity was created by a lack of infills at the ground floor.

Overturning forces from the wall above exceeded the strength of the columns designed for gravity loads only (Figure 41). Strengthening columns to resist gravity plus overturning forces is the minimum requirement for retrofitting the building. The shear transfer requirements and potential of the diaphragm at the level of discontinuity should also be checked.

Asymmetric distribution of mass and lateral stiffness causing torsion in plan caused considerable damage in many buildings. Torsional forces rotate floor diaphragms and displace vertical elements sideways. The ground floor columns in the first storey of one of the apartment complexes were subjected to lateral displacements due to torsional rotation of the first floor (Figure 42). The eccentricity between the centre of mass and the centre of rigidity can be reduced by designing more carefully positioned shear walls.

Buildings are recommended with a sufficient seismic gap to avoid pounding or hammering damage. When adjacent unseparated structural units are at different heights the lower building receives an unexpected loading while the higher building suffers damage at the roof level of the lower building. An isolation joint reduces pounding damage and permits both structures to move independently.

Walls with large openings for windows and doors reduce the in-plane shear strength of masonry walls. Diagonal shear cracking of masonry piers was common in the quake-affected area (Figure 43). A two-storey duplex apartment complex suffered heavy in-plane shear damage to its brick masonry walls (Figure 44). Poor building material and quality of construction also contributed to the collapse.

The dynamic stability of a masonry wall under out-of-plane loads is related to its height to thickness ratio. Most serious failures due to this deficiency occurred in staircase areas of buildings.

Decreasing the slenderness of top storey walls and anchoring the wall to the roof slab can help avoid out-of-plane failures. The slenderness ratio for single storey buildings should be less than 21, for multi-storey buildings less than 15, and for top storeys less than 9.

Figure 41.   Crushing of a RC column due to axial forces from the overturning and sway *moment.*

Figure 42.   Columns of open bays crushed by overturning forces from walls above.

Figure 43.   Shear cracking of masonry piers.

Figure 44.   Severely damaged duplex unit walls.

Earthquakes expose errors, poor workmanship, inferior quality building materials, and poor project coordination in construction work. Cheaply built residential quarters were badly affected.

6.3   *Building element performance*

Unreinforced brick infill walls crack diagonally at relatively small deformations (Figure 45). However, they increase the lateral strength of weak RC frames and provide a redundant load path to carry both vertical and horizontal forces to the

Figure 45.   Shear & flexural cracking of infill.

Figure 46.   Damage to parapet wall.

foundation. Brittle failure of unreinforced masonry infills was the most common structural damage in the affected area and resulted in the dangerous shedding of masonry.

There are several methods to remedy infill deficiencies. First, infills can be isolated from the frame by a sufficient gap to accommodate seismic deformations of the frame. Out-of-plane support must then be provided. If the frame alone is deficient in lateral strength, stiffness and ductility it may need to be strengthened. Secondly, infills can be treated as shear walls. Their inadequate in-plane shear capacities may need upgrading.

Damage to staircase enclosing walls at roof level was widespread. This can be attributed to large amplification of ground motion at upper floors and excessive slenderness of the walls. Stairwells can attract large seismic forces due to the bracing truss action provided by unseparated inclined members and they often cantilever above roof level and are located at a diaphragm discontinuity.

Damage to unreinforced parapet walls was common (Figure 46). Where the height of parapet walls exceeded twice their thickness walls should be provided with vertical reinforcement and anchored back to roof or floor.

The Indian Standards do not have specific provisions for the design of weak RC frame with brick infill construction, despite brick infills being the most popular element for RC frames.

Although the majority of buildings were not designed for seismic loads, many of them survived the earthquake sustaining only minor damage. One or two storey unreinforced masonry buildings performed satisfactorily, especially those that were well constructed and did not have major layout or planning deficiencies. Masonry infill walls in weak RC frames clearly enhanced the overall shear resistance of the structure when the integrity of RC frames themselves under lateral loads was doubtful.

The extent of damage would have been significantly reduced if modern earthquake-resistant design procedures and construction practices included in the Indian seismic codes had been followed.

The widespread damage to infill walls requires more in-depth study. Despite four decades of extensive research, a consensus is yet to emerge on the design, strengthening and ductility evaluation of infilled frames. There is an urgent need to develop techniques to increase the ductility of unreinforced infills.

7 SAFEGUARDING MASONRY WALLS

7.1 *Introduction*

Horizontal bands are provided in masonry buildings to improve their earthquake performance. These bands include plinth band, lintel band and roof band. Even if horizontal bands are provided, masonry buildings are weakened by the openings in their walls (Figure 47).

During earthquake shaking, the masonry walls get grouped into three sub-units, namely spandrel masonry, wall pier masonry and sill masonry. In the northern state of Jammu & Kashmir, it is normal to find wall panels interrupted by crossing timber elements that increase the integrity of the panel (Figure 48).

Consider a hipped roof building with two window openings and one door opening in a wall (Figure 49a). It has lintel and plinth bands. Since the roof is a hipped one, a roof band is also provided. When the ground shakes, the inertia force causes the small-sized masonry wall piers to disconnect from the masonry above and below.

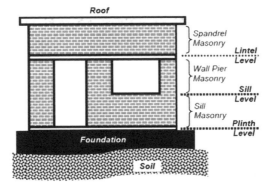

Figure 47.   Sub-units in masonry building – walls behave as discrete units during earthquakes.

Figure 48.   Wall panels interrupted by cross-timber elements.

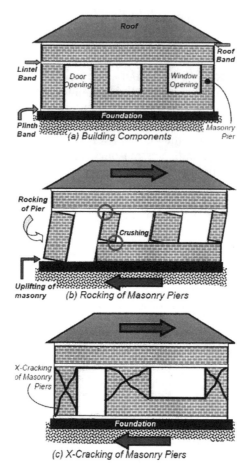

(a) Building Components

(b) Rocking of Masonry Piers

(c) X-Cracking of Masonry Piers

Figure 49.   Earthquake response of a hipped roof masonry building (no vertical reinforcement is provided in walls).

These masonry sub-units rock back and forth, developing contact only at the opposite diagonals (Figure 49b).

The rocking of a masonry pier can crush the masonry at the corners. Rocking is possible when masonry piers are slender, and when weight of the structure above is small. Otherwise, the piers are more likely to develop diagonal (X-type) shear cracking (Figure 49c); this is the most common failure type in masonry buildings.

In un-reinforced masonry buildings (Figure 50), the cross-section area of the masonry wall reduces at the opening. During strong earthquake shaking, the building may slide just under the roof, below the lintel band or at the sill level.

Sometimes, the building may also slide at the plinth level. The exact location of sliding depends on numerous factors including building weight, the earthquake-induced inertia force, the area of openings, and type of doorframes used.

Embedding vertical reinforcement bars in the edges of the wall piers and anchoring them in the foundation at the bottom and in the roof band at the top (Figure 51), forces the slender masonry piers to undergo bending instead of rocking. In wider

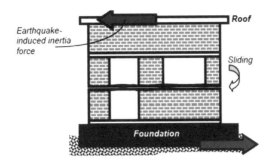

Figure 50. Horizontal sliding at sill level in a masonry building – no vertical reinforcement.

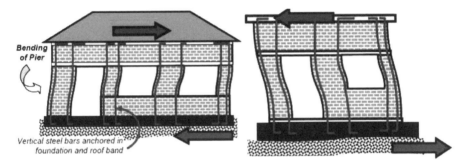

Figure 51. Vertical reinforcement in masonry walls-wall behaviour is modified.

wall piers, the vertical bars enhance their capability to resist horizontal earthquake forces and delay the X-cracking.

Adequate cross-sectional area of these vertical bars prevents the bar from yielding in tension. Further, the vertical bars also help protect the wall from sliding as well as from collapsing in the weak direction.

7.2 *Protection of openings in walls*

Sliding failure mentioned above is rare, even in unconfined masonry buildings. However, the most common damage, observed after an earthquake, is diagonal X-cracking of wall piers, and also inclined cracks at the corners of door and window openings.

When a wall with an opening deforms during earthquake shaking, the shape of the opening distorts and becomes more like a rhombus – two opposite corners move away and the other two come closer. Under this type of deformation, the corners that come closer develop cracks (Figure 52a).

The cracks are bigger when the opening sizes are larger. Steel bars provided in the wall masonry all around the openings restrict these cracks at the corners (Figure 52b). In summary, lintel and sill bands above and below openings, and vertical reinforcement adjacent to vertical edges, provide protection against this type of damage.

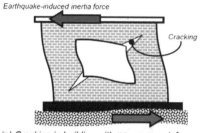
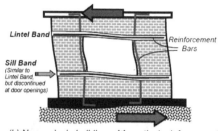

(a) Cracking in building with **no** corner reinforcement   (b) No cracks in building **with** vertical reinforcement

Figure 52.   Cracks at corners of openings in masonry building reinforcement around them help.

8   SOME SUGGESTIONS FOR EARTHQUAKE RESISTANT MASONRY

Since the brittle nature of masonry buildings is the major cause for collapse of buildings and loss of lives, there is a need to introduce remedial measures in the construction of such buildings.

The horizontal bands are helpful in tying the walls together at the junctions and also in preventing the growth of vertical cracks and in-plane shear cracks. However, they may not be adequate in strengthening against out-of-plane flexure, especially for flexure cracks that run horizontally. In this context the Department of Civil Engineering, Indian Institute of Science, Bangalore has developed the concept of "containment reinforcement" (Raghunath S et al 2000), to contain the flexural tensile cracks from growing. This has also helped in imparting ductility and in absorbing a lot of energy during earthquakes.

9   POST-TENSIONED MASONRY STRUCTURES

Post-tensioning offers a new potential to innovative engineers and architects for the revival of masonry as a structural material. Plenty of types of constructions are feasible at costs competitive with reinforced concrete structures. Figure 53 illustrates a selection of some of the most straightforward applications of post-tensioned masonry.

In residential and office buildings primarily walls in the upper storeys, would benefit from post-tensioning both for strength and in-service performance, Figure 53a. At lower storeys, gravity loads will reduce the required amount of post-tensioning, in general. Basement walls, subjected to out-of-plane lateral earth pressure, are another application in residential buildings, Figure 53b.

Post-tensioned masonry may be used to infill large frames in industrial buildings, Figure 53c. Apart from cast-in-place construction, post tensioning offers benefits to prefabricated walls during transport and erection and could be used to effectively connect the walls to cast-in-place elements, Figure 53d. Tilt-up masonry walls and sound walls seem to be other potential applications. Lots of

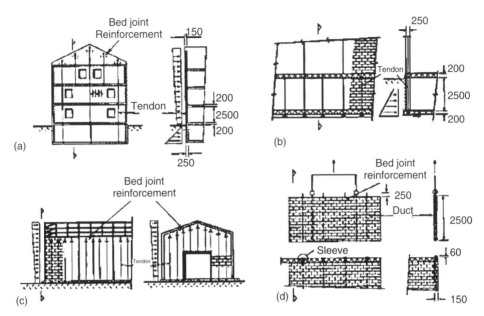

Figure 53. Typical future applications of post-tensioned masonry for new structures (a) Residential building; (b) Basement wall; (c) Infilled frames; (d) Prefabricated walls.

masonry constructions were built at a time when people were not yet as concerned as today about the strength for lateral wind and seismic loads. Such constructions, either individual walls or entire buildings, can be strengthened by post-tensioning to comply with the most recent requirements.

10   CONCLUSIONS

An effort has been made to bring out the diversity in the way people in India construct their homesteads and other buildings. Given the geographical extent of India and also the cultural variety one finds that every region has developed its own style corresponding to the materials available and also the environmental factors. But there is a certain level of uniformity also.

India suffers from natural disasters almost routinely. While events like flooding, cyclones are amenable to a level of warning, earthquakes are sudden and efforts at prediction are being shelved, as in Japan. Hence, to assure the safety of the public, the only recourse is to make the building the structures strong and safe.

With this in mind, the paper has devoted many words to studying the nature of weaknesses that the traditional structural forms of masonry enclosures and also how they could be avoided. It is hoped that the broad statements would spur interest and people would delve into the details and what results would be a strong and safe structure.

11   REFERENCES

[1] National Building Codes.
[2] CPWD Codes & Specifications.
[3] State PWD Codes & Specifications.
[4] Indian Standard Codes.
[5] Earthquake Engineering Research Institute – World Housing Encyclopedia Reports on:
   a. Rubble Stone Masonry Walls with Timber Frame and Timber Roof.
   b. Reinforced Concrete Frame Building with Masonry Infill Walls designed for Gravity Loads.
   c. Unreinforced Brick Masonry Building with Reinforced Concrete Roof Slab.
   d. Unreinforced Brick Masonry Walls in Mud Mortar with Flat Timber Roof.
   e. Unreinforced Brick Masonry Walls with Pitched Clay Tile Roof.
   f. Rural Mud House with Pitched Roof Traditional Rural House in Kutch Region of India.
   g. Low-Strength Dressed Stone Masonry Buildings.
[6] "Behaviour of masonry structures during the Bhuj earthquake of January 2001", by K. S. Jagadish, S. Raghunath and K. S. Nanjunda Rao.
[7] "Earthquake Tips from Indian Institute of Technology Kanpur", Building Materials and Technology Promotion Council, New Delhi, India.
[8] "Properties of Masonry, Design Considerations Post-Tensioning System for Masonry Structures Applications", VSL International Ltd. Berne, Switzerland.
[9] Earthquake Hazard Centre Newsletter: Jabalpur Earthquake, India, May 22, 1997: Reconnaissance Report – A summary from the report by Dr. Durgesh C. Rai, Dr. J. P. Narayan, Dr. Pankaj and Dr. Ashwani Kumar, September 1997, Department of Earthquake Engineering, University of Roorkee, India

CHAPTER 8

Typical masonry wall enclosures in Italy

Roberto CAPOZUCCA

Associate Professor
Facoltà di Ingegneria
Università Politecnica
delle Marche
Italy

SUMMARY

According to the strategy defined by the CIB Commission W023-Wall Structures, the Italian building masonry enclosure system is discussed in this paper. After an analysis of developments of the most common types of masonry and a description of the most frequent structures and masonry materials, the principal enclosure systems are presented. In particular, methods of calculation in the case of seismic situation are described.

1 INTRODUCTION

The aim of this paper is to present the state of the art of the Italian practice for the use of masonry enclosure systems, according to the strategy defined by the CIB Commission W023-Wall Structures.

This masonry wall system is recent and is linked to a different construction method with respect to those used in historic masonry buildings. The enclosure masonry wall system was developed following the increased use of RC and steel structural frames in buildings.

Before analysing the enclosure systems, it is useful to investigate the evolution of masonry walls, which in Italian regions is extremely complex, in the majority of masonry structures and materials. In fact, a characteristic of masonry is the large variety of systems used for the masonry walls linked to the material available in the region. Italy, for instance, is a country with significant differences from one region to another, so that each practical experience is often very different.

In this paper, the author presents a brief analysis of the evolution of Italian masonry system considering the materials that have been used from historical masonry up to the most recent developments in masonry structures and enclosure systems.

Particular attention is given to the problem of the seismic behaviour of buildings with RC frames with masonry enclosures, taking into account the Italian rules of practice.

To begin with, the Italian economic trends in the building sector, which plays an important role in national economy, is described through the data of the building associations.

2   BUILDING SECTOR AND MASONRY PRODUCTION IN ITALY

According to recent ANCE (*Associazione Nazionale Costruttori Edili*) forecasts, building investments in Italy in 2002 increased by 2.3% with respect to 2001 [1]. This upward trend in demand is expected to continue, be it at a lower rate, throughout 2003.

Table 1 shows the percentage trends in building investments in Italy (final for the year 2002 and forecast for 2003).

In particular, investments for residential buildings increased by 2.5% in 2002 due to the increase in new homes (2.0%) and restoration works (3.0%). The latter market segment seems to be the only segment experiencing an acceleration of the growth rate.

The brick production sector obviously follows the trends of the building sector and is currently benefiting from this upward trend although a slack, and even an inversion of this positive trend is expected in the years ahead. Total production increased by 3.5% in 2002 versus 2001, reaching a production volume of more than 18.7 million ton (Table 2).

Production is on the whole increasing, although there are consideration internal differences. The production of full and bricks with holes continues to decrease compensated by the upward trend in blocks for load bearing structures.

Single-block walls (Figure 1) consisting of single-layer vertical structures offering both resistance and thermal-acoustic insulation are consequently used with major frequency.

Table 1.   Building investments – Percentage variations (final figures for 2002 and forecast for 2003).

(*)Type of building	2001	2002	2003
Buildings	3.7	2.3	1.6
Homes	3.0	2.5	1.6
New	3.6	2.0	1.2
Restoration works	2.5	3.0	2.0
Other buildings	4.5	2.1	1.6
Non residential buildings	5.9	2.8	1.5
Public works	2.5	1.0	1.8

(*) ANCE economic observatory.

Table 2. Brick production in Italy.

(*)Type of product	2001 (ton 3 105)	2002 (ton 3 105)	%
Full bricks	13.5	11.6	−13.8
Hollow bricks	40.8	42.1	3.2
Blocks for load bearing masonry walls (normal brick)	17.1	19.7	15.1
Blocks for load bearing masonry walls (lightened brick)	23.3	25.7	10.6
Enclosure blocks (normal brick)	5.3	5.7	7.3
Enclosure blocks (lightweight brick)	7.5	6.2	−18.1

(*) ANCE economic observatory.

Figure 1. Modern structural masonry with one-block wall. [2]

Single-layer wall systems are also used for non load bearing masonry structures, known as wall enclosure systems.

Finally, the masonry building sector has also been considerably shaken by the recent publication of the new seismic law issued by G.U. 105 dated 8th May 2003, according to which the entire territory is classified as seismic, with varying degrees of seismic risk.

As things stand therefore, bricks with holes coverage of between 45%–55% can no longer be used for load bearing structures and may be used only as bricks in wall enclosure systems.

Substantially, the brick market on the whole seems to be experiencing a period of expansion in terms of sales volumes, even if the general trend seems to consist of products with a decidedly upward trend and others with a downward or at least irregular trend.

3 EVOLUTION OF THE ITALIAN MASONRY WALLS

3.1 *Historic masonry walls*

Traditional walls generally known more simply as *masonry* date back to very ancient times and, despite the fact that they were *unfaced*, profoundly affect the formal architectural style to the extent that, often, when examining the many materials making up a masonry structure, it is often difficult to make a distinction between the functional and aesthetical structures.

One of the most ancient forms of masonry is cyclopean or boulder masonry, which consists of enormous irregular boulders laid together without mortar along carefully finished surfaces. This masonry technique was used to build the Etruscan towns in central Italy, although cyclopean stone masonry was used above all for fortifications, supporting structures or very large buildings.

The most ancient masonry technique which more than any other resembles modern building techniques is that introduced by the ancient Romans who used a two-layer system: parallelepiped blocks laid with the end facing the wall and other blocks laid with the end orthogonal to the wall (Figure 2).

In the *opus quadratum* system, horizontal courses or rows and horizontal joints are staggered. A wall of this kind was built according to specific geometric rules, both in terms of the preparation of the elements making up the wall and their position.

Opus quadratum masonry was used by the ancient Romans for their most important buildings. It is found in monumental buildings and on the ground floor of structures rebuilt on ancient ruins. The mortar in the thin joints, was generally sufficient to assure the contact between the blocks. The longitudinal units (*ortostrati*) are alternated to units (*diatoni*) linking the two faces.

Clay bricks, made by pressing the clay into moulds, were also commonly used for the construction of ancient buildings. Another form of masonry originating as poor raw clay masonry is concretion masonry which was widely used by the ancient Romans. Concretion masonry was made by laying mortar and stone between two outer facings.

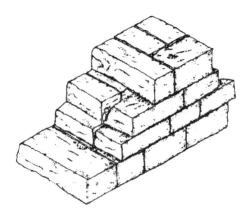

Figure 2. Historic Roman masonry opus quadratum.

While in the case of squared stones walls or *opus quadratum*, no mortar is required because its solidity was ensured by the skilled arrangement of the stones, adhesive material is essential in the case of walls made with the concretion system.

In the Middle Ages, masonry with intermediate characteristics was used with respect to that of ancient times: prevalently large irregular stone blocks were laid with mortar and smaller fragments, to form an external texture which was just as mechanically efficient as that of the classical structures, containing material of lower quality, often even without mortar.

This type of masonry is known as masonry with two leafs (*muratura a sacco*). This type of masonry was used until the XIX century replacing stone facings with a combination of stone and brick or brick facings.

The external face is organised with stones and/or clay bricks. The arrangement has regular courses of different heights, in that the junctions between elements are much thinner that the pebble stone wall. The external face is often a quarry face. The internal face, on the other hand, is made of rubble, while smaller stones and stone flakes are inserted in the joints. The internal bag sometimes is absent (double-leaf wall). The structural strength depends on the large elements. Often there are no *diatoni*, even for the stone lengths which are difficult to find and lay, like the thickness.

In other cases, mixed masonry consisting of pebbles, rubble stone and bricks joined with lime mortar, and brick facings are present in historic masonry. There are two different types of facings: the external face is made of bricks laid longitudinally alternated with bricks that join the other face. The thickness is generally approx. 50–60 cm. This dimension is adequate for load-bearing walls. It is in any case necessary to verify the presence of elements passing through the thickness.

Examples of buildings constructed with the *muratura a sacco* system, with sandstone and brick facing, are shown by Figure 3.

Figure 3.   Mediaeval buildings built with the muratura a sacco system. [3]

In Italy, historical centres are still relatively intact with mediaeval and post-mediaeval masonry where the *muratura a sacco* technical undoubtedly prevails. Unfortunately this type of masonry has often proven to be very sensitive to earthquakes.

3.2   *Typical Italian buildings with structural masonry until XX century*

Following the advent of the industrialised production of bricks, the masonry systems used have in same way returned to the ancient *opus quadratum* system, replacing the stone units with bricks.

Figures 4a and 4b show traditional "gothic" type two and three head masonry. In each layer two bricks (Figure 4b) are laid longitudinally and alternated with a brick laid transversally. In the upper logging the transversal brick is laid in the middle of the longitudinal brick, to obtain the logging. The next two courses are aligned to these respectively. This type of masonry arrangement is one of the best in terms of both static and seismic resistance.

Civil buildings constructed with load bearing solid brick walls represent the most commonly used technique for homes built in the XIX century through to the

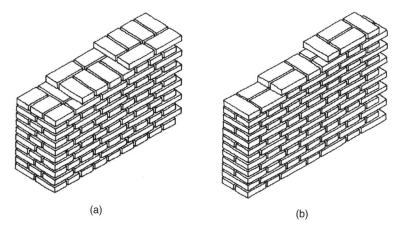

(a)   (b)

Figure 4.   Gothic masonry with full two (a) and three (b) head bricks.

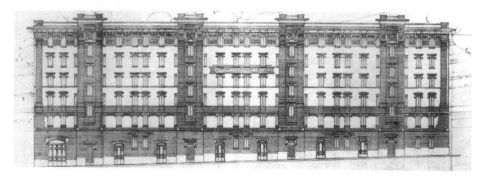

Figure 5.   Project by Q. Pirani (1912) for a load bearing masonry building.

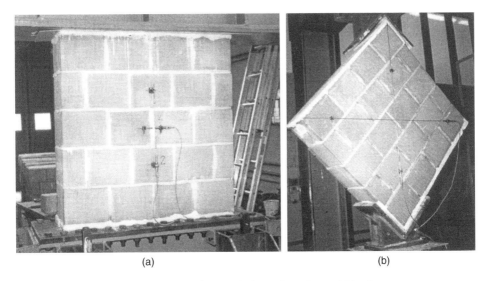

(a)   (b)

Figure 6.   Modern masonry panels consisting of perforated blocks.

early decades of the XX century with load bearing floors in wood and iron, until RC slabs were introduced in the early 900's. Figure 5 shows a building project (Q. Pirani, 1912) in Rome [4].

In recent decades, after the post-war period during which the *opus quadratum* system had been almost totally abandoned, there has been renewed interest in buildings with load bearing structures, leading to the production of new types of bricks.

The use of perforated blocks for the construction of load bearing walls has become particularly popular.

Figure 6 shows blocks subjected to the vertical and diagonal compression tests commonly used in the construction of modern buildings with load bearing walls [5].

3.3   *Enclosure wall system in modern buildings with RC or steel frame*

Non-load bearing walls used to face RC or steel load bearing structures are in some way an evolution of traditional masonry. Like the latter in fact, they have the by no means secondary function of delimiting the architectural space and thermally insulating the rooms of the home, without having a pre-eminent load bearing function.

Enclosure walls are used to close the external surfaces of the resistant frame of the building. Their conformation depends on the structure above ground, namely whether this is in reinforced concrete and steel and on the functional requirements of the building.

Contrary to traditional masonry buildings, for which good resistance is required, in the case of enclosure walls whose function is exclusively to protect the internal spaces from weather, heat fluctuations and sound transmission, other properties are required such major lightness, high thermal and sound insulation properties, adequate durability and low costs.

Although there are some interesting examples of enclosure walls in previous historical periods, such as the ancient wooden frame structures of the XVIII century in

Southern Italy used to build earthquake resistant houses (an interesting example of which is the building project by G. Vivenzio, 1783, characterised by resistant wooden trusses [6]). The introduction of enclosures in the modern age is relatively recent.

This technique is tied to the widespread use of independent frame structures, the new industrialisation processes and building systems.

In Italy, according to the type of resistant frame and construction system used, there are generally two types of enclosure walls:

(a) actual enclosure walls with the wall panel incorporated in the frame, made of materials having appropriate technological characteristics that are laid with procedures not very different to those used for traditional masonry walls (Figure 7);
(b) prefabricated enclosure panels which are applied to the facing to cover the structural frame. These must be fitted with assembly means and require special fixing means on the resistant load bearing structure (Figure 8).

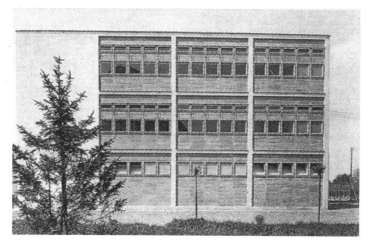

Figure 7.   Building for medical laboratory by I. Gardella (1939). [7]

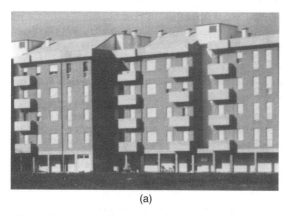

(a)

(b)

Figure 8.   Example of a building with RC panel enclosure. [8]

4 MATERIALS FOR ENCLOSURE MASONRY WALLS

4.1 *Clay units using in masonry enclosure walls*

Figure 9 shows a typical cross-section of an enclosure wall known as "*muro a cassetta*" which represents the most commonly used building technique for enclosure wall panels applied to RC frames for buildings. This is a double wall with back up filled with insulating material such as polystyrene or glass or mineral wool.

Figure 9 shows some of the most common thickness of the structural elements [9]. Figure 10 shows some of the bricks commonly used for the thick section of the wall – the external enclosure "box" wall and relevant dimensions. This wall is generally built with hollow bricks (Figure 10a) or blocks (Figures 10b and 10c).

More recently the tendency is to eliminate the double layer wall known as "*muratura a cassetta*" in an attempt to obtain the same results by using thicker brick blocks to make a single enclosure wall (Figure 12).

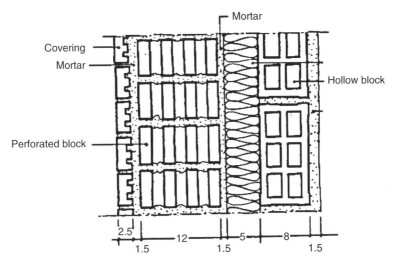

Figure 9. Typical cross-section of an enclosure wall.

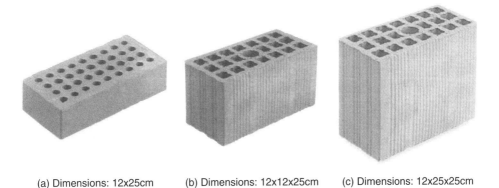

(a) Dimensions: 12x25cm   (b) Dimensions: 12x12x25cm   (c) Dimensions: 12x25x25cm

Figure 10. Brick and perforated blocks used for the outer leaf of the enclosure walls.

Dimensions 12x12x25cm

Figure 11.   Typical hollow block used for the internal leaf of the enclosure wall.

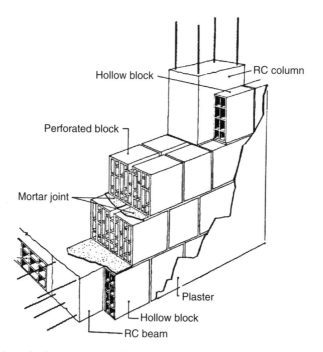

Figure 12.   A typical masonry enclosure system with only one unit. [9]

These are blocks with a height of no more than 20 cm and holes coverage between 50%–60% of the gross cross-section, with vertical grooves to facilitate wedging, doing away with the need for vertical mortar joints.

The mortars used for enclosure walls designed for the traditional double-leaf walls known as are mortars of a fairly low quality containing lime.

For block walls, pre-packaged mortars are used, generally with characteristics similar to type M2 as provided by Italian code of practice [10]. Figure 13 shows a number of perforated blocks commonly used for this type of enclosure wall with their dimensions.

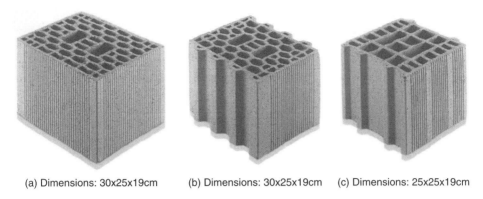

(a) Dimensions: 30x25x19cm   (b) Dimensions: 30x25x19cm   (c) Dimensions: 25x25x19cm

Figure 13.   Brick block units used for enclosures.

Figure 14.   Prefabricated panel in reinforced masonry. [8]

4.2 *Prefabricated enclosure panels with reinforced masonry*

A technique developed above all in the 80's is the reinforced brick wall technique with insulating layer and an internal lightened concrete surface (Figure 14).

For several years these panels also represented the construction technique used for buildings with prefabricated panel load bearing structure.

5 INFLUENCE OF THE ENCLOSURE WALL ON THE SEISMIC RESPONSE OF BUILDINGS WITH RC STRUCTURES

5.1 *Introduction*

A number of authors [11] have pointed out the very significant effect that masonry enclosure walls and internal masonry panels may have on the actual behaviour under lateral loads of a RC framed structure during an earthquake. Below briefly it refers to a problem that requires in-depth investigation and for which appropriate evaluation criteria are as yet unavailable.

Some experimental studies on this subject have thrown light on the stiffening and strengthening effect these elements exert when effectively inserted into a grid element of a reinforced concrete frame.

This effect, under not especially intense horizontal forces, could be simulated by considering an equivalent strut of thickness t assessable on the basis of the stiffness ratios of the two coupled structures: frame and closure wall.

The equivalent strut responds well in the elastic phase in terms of evaluating the enclosure's influence on the system stiffness, namely, for the calculation of the horizontal displacements. It can therefore be introduced into the frame for the purpose of calculating the distribution of the horizontal forces over the various elements of a structural complex. The equivalent-strut scheme does not however reproduce very well the other elements of the cladding walls.

Principally and synthetically, these factors are the following [11]:

– The equivalent strut does not reproduce the constraint that the closures exert on the rotations of the frame nodes. If next to a column of cladding elements there is another, the scheme does not let the effective rigidity of the joint, to which the adjacent beams are subject, be evaluated. This difficulty could be partially overcome by considering a fixed-joint constraint of the strut itself.
– The coupling, frame plus wall, regards two elements, the second of which displays a stiffer and more fragile behaviour, especially when the cladding walls are in hollow blocks with more than 40% of holes. This means that, if the structural element is subjected to strong lateral strains, after an initial phase in which the cladding wall shows itself very effective, it may then be destroyed and therefore does not collaborate any more in the subsequent phases of the earthquake. The partial advantage of the energy that has been dissipated by its destruction remains however, and this can be important to the overall behaviour of the structural complex.
– The cladding walls offer very low resistance to forces acting perpendicularly toothier plane. They can easily be destroyed by these forces, and it is not easy to take construction measures that avoid this.

The last two points definitely throw into relief the scant reliability which can be given to cladding walls, regarding factors of not easily solved problems. Anyway, their considerable stiffening, strengthening and dissipative capacities are very high.

Examination of the behaviour of buildings in areas struck by the most severe earthquakes makes it easy to realize how they have contributed in a determining manner to save from collapse a considerable number of buildings; and many other times they have anyway limited the consequences of more serious damage.

Therefore to renounce the benefit of their contribution can not be an appropriate policy. More appropriate instead would be the criterion of intervening, at least at the qualitative level, during the phase of architectural design of the building, by taking care to provide the most rational distribution of these elements.

This operation should be performed in the spirit of seeking the general benefit that the cladding walls are able to furnish, and that is certainly high even if not easily quantifiable. But those distributions that can give rise to the well known negative effects should be avoided; in any event these negative distributions are briefly brought to mind in a following paragraph.

As regards construction methods, it is important that the closures be composed of light but not fragile bricks, and that the maximum bonding with the mortar be affected. Therefore hollow brick (vertical holes, no greater 40%) or lightened concrete blocks are to preferred. Care must then be taken that the last layer of mortar be well executed, so as to obtain the best possible connection with the beam of the framework.

5.2 *Negative effects of closure walls on the behaviour of building with RC frames*

It has been seen how important it is that the building collapse mechanism not derive from the formation of a soft storey. But it often happens that current architectural formulations tend, for various motives, to create so-called *pilotis*-type configurations, thus favouring the formation of this dangerous mechanism (Figure 15).

5.3 *Calculation of enclosure masonry walls by the Italian Code of Practice*

The Italian Code of Practice [12] for buildings in seismic zones provides a method of calculation for the analysis of RC frames or steel frames with masonry panels.

Figure 15.   Effects of a masonry enclosure in a building with RC frame. [11]

The calculation procedure indicated permits reproducing with sufficient approximation the behaviour of a frame with masonry enclosure wall subject to lateral forces, provided the following conditions are satisfied:

– the frame must consist of RC or steel elements suitably connected at the nodes and must adhere to the enclosure. The enclosure wall must be efficiently connected to the frame so as to ensure contact and adherence, thus guaranteeing the transmission of normal and shearing forces;
– the ratio between the sides of the wall panels must generally be between 0.5 and 2.0;
– the slenderness of the panel must not be above 20;
– the enclosure panel must not have opening other than those delimited by RC frames ensuring continuity.

In terms of evaluating the behaviour of the frame-enclosure system and sharing the horizontal seismic forces between the resistant elements, the effect of the enclosure may be taken into account considering the performance of an equivalent diagonal truss (Figure 16).

The truss must have thickness t of the masonry and width s equal to 1/10th of the length of the diagonal. An equivalent system can thus be considered consisting of frame beams and columns and of the above diagonal trusses deemed to be hinged at the ends (Figure 17).

The stiffness of each truss will be equal to:

$$(EA/d)_{eq} = 0.1 \cdot E_m \cdot t \tag{1}$$

where, $d = (l^2 + h^2)^{1/2}$ is the length of the diagonal, while E_m is the elasticity module of the masonry.

The lateral behaviour of a frame is strongly affected by the interaction produced by the enclosures on the floor of the frame itself, provided these are efficiently connected. In particular, according to Italian code, leaf enclosure walls or those with an enclosure above 45%, can not be considered.

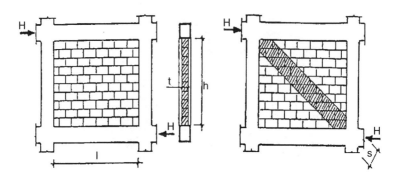

Figure 16.   Equivalent masonry truss for the enclosure panel.

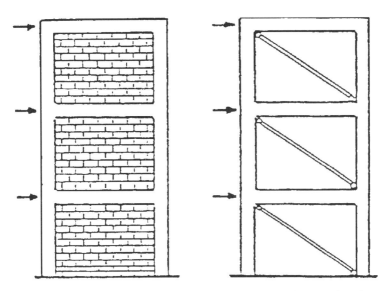

Figure 17.   RC frames with masonry enclosures and equivalent static scheme.

With reference to a RC frame interacting with an incorporated wall panel with thickness t, length l and height h, Italian Code provides three wall breakage mechanisms:

(a) failure by horizontal force due to the tangential stresses acting on the central area of the enclosure;
(b) diagonal failure by tension due to shear acting on the central area of the enclosure;
(c) failure by local crushing of the enclosure edges due to the concentration of interaction forces transmitted by the frame.

The verification of resistance for the three breakage conditions are carried out on the basis of the following relationships:

– failure by horizontal shear stress:

$$H_0 \leq \left[\frac{\tau_u}{\phi}\right] \cdot 1 \cdot t \tag{2}$$

$$\tau_u = f_{vk0}\sqrt{1 + \frac{\left(0.8 \cdot \dfrac{h}{l} - 0.2\right)}{1.5 \cdot f_{vk0}} \cdot \frac{H_0}{1 \cdot t}};$$

where:

– diagonal failure:

$$H_0 \leq \left(\frac{f_{vk0}}{0.6 \cdot \phi}\right) \cdot 1 \cdot t \tag{3}$$

– crushing of the edges:

$$H_0 \leq 0.8 \cdot \frac{f_k}{\phi} \cdot \cos^2 \theta \cdot \sqrt[4]{\frac{E_c \cdot I \cdot h \cdot t^3}{E_m}} \tag{4}$$

where H_0 is the horizontal seismic force acting on the masonry element – namely the horizontal element of the force acting on the equivalent truss – evaluated according to the structure coefficient introduced by Italian Code of Practice; f_{vk0} characteristic shear strength of the masonry without vertical loads; f_k typical compressive resistance of masonry; θ angle of the diagonal of the panel with respect to the horizontal; ϕ the reduction factor of the tensions; E_c and E_m respectively, the elasticity modules of concrete and masonry; I moment of inertia of the transverse cross-section of the RC column.

Finally, it provides a series of additional controls on the RC structures of the frame:

– assessment of the axial forces in the columns;
– shear assessment of the RC column taking into account the increase of force H_0 calculated for the truss equivalent to the enclosure in the static scheme;
– bending assessment with a bending moment obtained from the frame static scheme with equivalent truss increased by the value equal to $M = \pm H_0 \cdot h/10$.

6   CONCLUSIONS

The masonry enclosure system is a recent method of construction for masonry walls. The enclosure masonry wall system was developed following in Italy the increase of use of RC and steel structural frames in building after the Second World war.

In the paper, before analysing the enclosure systems, it was investigated the evolution of masonry walls in Italian regions where the masonry is extremely different, in the majority of masonry structures and materials.

In the recent years the enclosure walls are built by perforated blocks even if the typical method of construction of enclosure walls is represented by the so called *muro a cassetta*. In this system two masonry leafs are separated and between them insulator material is disposed.

The most important problem in the structural behaviour of buildings with enclosure walls in seismic area is the interaction with frames. The Italian Code of Practice permits to analyse the RC and steel frames with the presence of diagonal trusses characterised by an equivalent stiffness for enclosure walls.

7   REFERENCES

[1] Ceriachi, C. – "L'industria italiana dei laterizi. Indagine conoscitiva sulla produzione 2002". L'industria dei Laterizi, A.N.D.I.L., Luglio-Agosto 2003, 229–235, Roma (in Italian).

[2] Holker, M. – "Blocchi rettificati: aspetti da considerare nella produzione e nell'-impiego". L'industria dei Laterizi, A.N.D.I.L., Settembre-Ottobre 1999, 321–324, Roma (in Italian).

[3] Capozucca, R.; Pellegrini, A.; Scolastra, A. – "Historical Masonry Shear Panels in Seismic Zone". VII Int. Seminar Structural Masonry, 2002, pp. 325–332, Brazil.

[4] Toschi, L. – "Quadrio Pirani e la casa popolare a Roma (1904–1914)". Costruire in Laterizio, No. 26, 1992, 118–127, Italy (in Italian).

[5] Capozucca, R. – "Analysis of Block Masonry Panels with Different Mortar Joints under Compression and Shear". VII Int. Seminar Structural Masonry, 2002, 177–183, Brazil.

[6] Tobriner, S. – "La casa baraccata: un sistema antisismico nella Calabria del XVIII secolo". Costruire in Laterizio, No. 56, 1997, 110–115, Italy (in Italian).

[7] Campioli, A. – "Alle origini del solaio in latero-cemento". Costruire in Laterizio, No. 26, 1992, 148–151, Italy (in Italian).

[8] Davoli, S. – "Lo stabilimento di Correggio (RE) della Unicoop". La Prefabbricazione, No. 2, 1982, 105–112, Milano (in Italian).

[9] Tubi, N. – "La realizzazione di murature in laterizio". ANDIL, Ed. Laterconsult, 1993, Roma (in Italian).

[10] D.M. 20 Nov. 1987 – "Norme tecniche per la progettazione, esecuzione e collaudo degli edifici in muratura e per il loro consolidamento" (in Italian).

[11] Parducci, A. – "Aspetti progettuali delle costruzioni antisismiche". L'Industria Italiana del Cemento No. 10, 1981, 749–778, Roma (in Italian).

[12] D.M. 1996 – "Norme tecniche per le costruzioni in zona sismiche" (in Italian).

CHAPTER 9

Enclosure walls in the Nordic countries

Tor-Ulf WECK

Professor
Helsinki University of Technology
Helsinki
Finland

SUMMARY

A survey of material percentages is in different structural purposes is given for Finland. The most common enclosure wall types in the Nordic countries are presented. In addition some newer, for time being, less common structures are presented.

1 INTRODUCTION

All Nordic countries had earlier a lot of forests. This naturally led to buildings being erected of wood. Originally houses were all log houses, but the development of mechanical wood technology led to timber framed houses with wood flakes as thermal insulation. This kind of timber-framed houses was typical for up to two storeys even in early 60's.

Higher buildings had the lower stories made of clay bricks. Multi-storey apartment buildings as well as office and other public buildings were made of clay bricks, too. These kinds of buildings were dominant until the early 1950's. The concrete walls were insulated with lightweight aggregate concrete with rendering protection as the outer façades.

Late 1960's saw the introduction of concrete element walls, especially in Finland and Sweden. Denmark was keener to keep the traditional clay brick walls and smaller scale housing being traditionally closer to the type used in Central European countries and especially North Germany.

During the last years some other types of wall enclosures are being used although the main enclosure in multi-storey apartment houses still is the concrete element structure.

Office buildings and especially industrial buildings have recently more often steel or glass facade. One family and detached houses are still often timber framed, but the outer façade made of clay bricks thus giving a more stable and expensive look.

2   BUILDING SECTOR

2.1   *Importance of the building sector*

The population of Finland is slightly above 5 million inhabitants. The construction activity represents roughly 10% of the Gross Domestic Product. The buildings are the most important activity.

The value of existing buildings is around 250 million euros, which is about half of the whole national capital. There are 60–70 square meters of buildings for each citizen.

The building activity follows the general economic trend, being one of the first areas to notice a change. At present, 2005, the general activity of the office building sector is declining, after several years of fast economic growth, but building of dwelling houses continues on a high level one reason being the low interest rate.

2.2   *Climatic effects on building*

Finland and other Nordic countries being in north, have the influence of frost to be taken into account in all circumstances where buildings are imposed to climatic conditions.

The effect of frost affects in different ways depending on the distance from the coast. Finland being the easternmost of the Nordic countries has partly a continental climate especially in the eastern parts of the country.

The Nordic situations requires also rather thick thermal insulation layer in outer walls.

A typical thickness of mineral wool is 175 mm and is increasing with tightening thermal insulation requirements. Thus tying together the outer and inner leaves of the wall requires special consideration due to the large cavity space for insulation.

Building in winter time favours building of prefabricated elements, too.

3   SOME STATISTICS ON USE OF BUILDING MATERIALS

The development of relative share of different building materials in load bearing structures is shown in Figure 1 below.

The figure gives somewhat wrong impression as the small housing is dominating. If only multi storey buildings are taken into account the picture is different as shown in Figure 2. The numbers on the columns are the percentages of that material of the total number of houses.

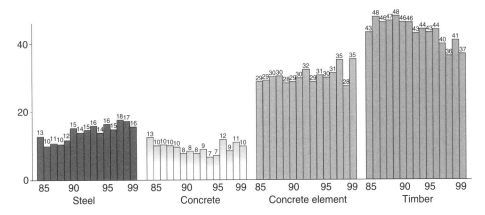

Figure 1.   Load bearing building material in all house buildings in Finland.

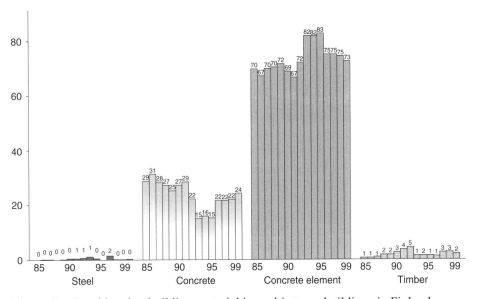

Figure 2.   Load bearing building material in multi storey buildings in Finland.

4   MOST COMMON ENCLOSURE WALL SYSTEMS

4.1   *Different types of enclosure walls*

Façade materials are in a much greater variety than materials for the load bearing structure. Traditionally the façade material in small houses is wood and in bigger buildings masonry structure made of clay bricks.

Although the main load bearing material in the latter buildings, nowadays, is concrete, the clay brick has prevailed its position as an important facade material.

The development of relative share of different building materials as facade is shown in Figure 3 below. One has to observe that clay brick covered concrete elements are considered as clay facades. The portion of this kind of clay facades is roughly 1/4 of all brick facades.

Taking into account only apartment buildings the figure is again different (Figure 4). In this case, the portion of brick covered concrete elements is roughly 1/2 of all brick facades.

Taking into account only small one and two storey buildings the present portion of materials load bearing structure is shown in Figure 5 and the portion of facade materials in Figure 6. As one can see from Figure 5, the timber framed house is the dominant structure. It can be made either by wood or consist of a masonry veneer wall.

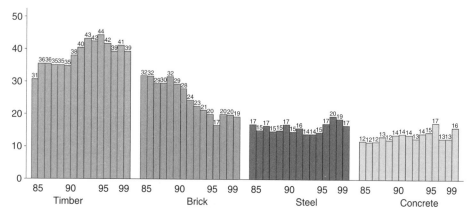

Figure 3.   Facade material in all housing buildings in Finland.

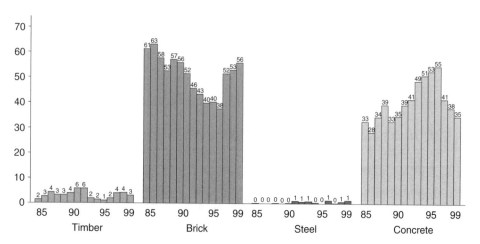

Figure 4.   Facade material in apartment buildings in Finland.

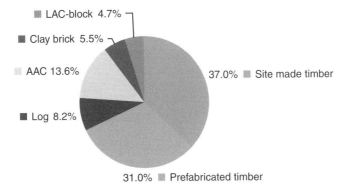

Figure 5. Material distribution in load bearing structure of small scale housing (%).

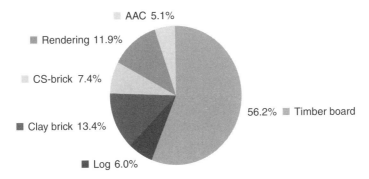

Figure 6. Material distribution in facades of small scale housing.

4.2 *Thermal insulation*

Thermal insulation in a Nordic country like Finland plays an important role in the enclosure wall. The thickness varies from 172 mm up to 250 mm. The most common type of insulation is mineral wool.

As described in Figures 7 to 9, there is a ventilation gap outside the insulation layer before the facing layer due to moisture penetrated through the wall especially during the cold season. In masonry structures this ventilation has lately been increased to 40 mm due to mortar falling to the cavity and in narrower gaps preventing effective ventilation.

4.3 *Damp proof courses*

A damp proof course is place between the wall and the basement to prevent the capillary water from penetrating into the wall. This is normally made either of thick plastic layer or bitumen sheet.

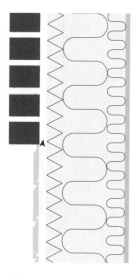

Figure 7. Timber framed structure.

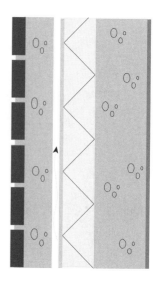

Figure 8. Sandwich element structure.

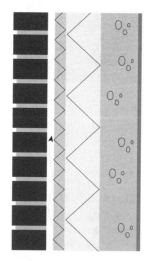

Figure 9. Masonry veneer wall structure.

4.4 *Wall finishes*

The most common facade materials are described in Section 3.1 and Figure 3.

The finishing is usually either the material itself or a paint laid on the material. Rendering is not very common finishing today.

Due to climatic conditions the rendering has to be well done and be made with frost resistant materials with good bond to the backing material.

4.5   *Most common wall enclosures*

Some common enclosure wall structures are described below.

The earlier mentioned timber framed house has a wall structure shown in Figure 7. In the figure, both masonry and timber, the most common facade materials have been presented.

The wall behind is in both cases similar, a timber frame with thermal insulation, the thickness of which is usually around 175–200 mm in order to fulfil the new Finnish thermal insulation requirements. If masonry is used, the thickness of bricks is usually 85 mm.

Multi storey apartment houses have usually outer wall made of sandwich element having outer surface made of concrete element that can have a slice of brick or tiles as the outer surface. This structure is shown in Figure 8.

The surface can also consist of coloured concrete or washed concrete.

Recently, a site made masonry veneer wall has, in some cases, replaced sandwich structure.

This is especially the case in public buildings and other buildings where a more living facade is the goal. Example of such structure is shown in Figure 9.

4.6   *Existing problems*

Walls described above are more or less a result of gradual evolution towards the most cost effective and in the Finnish climate long lasting enclosure walls.

Attempts to improve have been made during years and those successful have been incorporated into the building market and others have disappeared.

An effect caused by the authorities has been mainly on the thickness of the insulation layer, which is governed by the thermal insulation requirements, the base of which is the energy saving policy. Year 2003 saw the last tightening of the requirements.

5   TRENDS OF EVOLUTION

5.1   *Trends regarding the development of new enclosure wall types*

The ever-increasing cost of labour has resulted in enclosure wall types, which are fast to erect, partly factory made and longer lasting than the existing ones. At the same time they should be cost effective.

As complicated structures have greater possibility for erroneous workmanship, simplicity is one of the driving forces behind the development of new enclosure wall systems.

A possibility to build economically even during the winter period has led to industrialised building production where a great deal of the work can be done inside in controlled climatic conditions.

5.2   *New developments*

Figures 10 to 12 show photos of some structures that are not yet very common, but it can be expected that these structures will be more common during the coming years.

Figure 10.   Block masonry structure.

Figure 11.   Steel framed structure.

Figure 12.   Natural stone veneer wall structure.

The insulated block in Figure 10 is mainly used in low rise housing due to its simplicity. There is no separate installing of thermal insulation.

During the first years of the new millennium this type of block wall has increased its popularity.

A newer load bearing structure is steel frame houses replacing the timber ones. This is shown in a photo from an actual building site (Figure 11).

The structure is built in a similar way as the timber frame one.

Openings in the steel columns enhance the thermal insulation properties of the structure.

The natural stone facade in is use mainly in office and other buildings where an impressive outlook is sought (Figure 12).

The stone is fastened with mechanical anchors and not by mortar. This is due to temperature movement, which is different in natural stone and in the backing material, which usually is concrete.

In addition to the above mentioned structures there are numerous other structures, but none of them has gained a position, where they can be considered as gaining a large portion of the market.

6   SIGNIFICANT DOCUMENTATION

[1] Miettinen, Esko "Metallijulkisivut arkkitehtuurissa" (Metal facades in Architecture); Rakennustieto Oy Rati 2004; 176 p; ISBN 951-682-733-0 (in Finish).
[2] STEP 2 "Puurakenteet – rakennedetaljit – rakenteet" (Timber structures – details – structures); Rakennustieto Oy Rati 1998; 412 p; ISBN 951-682-401-3 (in Finish).
[3] Kinnunen, Jukka "Muuratut rakenteet – Rakennesuunnittelu" (Masonry structures – Design). Rakennustieto Oy Rati 2000; 158 p; ISBN 951-682-381-5A (in Finish).
[4] Miettinen, Esko & Ripatti, Harri & Saarni, Risto "Use of Steel in Housing Renovation". Rakennustieto Oy Rati 1998; 133 p; ISBN 951-682-483-8.
[5] Selonen, Olavi & Suominen, Veli "Nordic Stone". Unesco Publishing 2003; 64 p; ISBN 92-3-103899-0.
[6] Sundell, Kari (editor) "Kevytsoraharkot – Suunnittelu ja rakentaminen" (Lightweight aggregate blocks – Design and construction). Suomen Betonitieto Oy 2001; 125 p; ISBN 952-5075-37-0 (in Finish).

CHAPTER 10

Typical masonry wall enclosures in Poland

Jan KUBICA

Associate Professor
Dep. Building Structures
Silesian University of Technology
Gliwice
Poland

SUMMARY

The Polish building masonry enclosure system was strongly changed during last century. For a better understanding of the Polish reality on that field, a short review and discussion of the process of Polish masonry buildings systems evolution is presented and described in this paper. These changes and progress of masonry industry and practice were connected with increasing of requirements concerning with thermal protection of the buildings. Moreover, the main problems related with specificity of buildings situated on terrains subjected to static and dynamic influences connected with mining activity (in some industrial regions in South Poland) are also presented and discussed.

1 INTRODUCTION

The main aim of this paper is to present the changes and development of Polish masonry buildings practice from the end of the Second World War up to nowadays, with a special consideration to the present-day situation. It is fully compatible with the strategy defined on the CIB Commission W023-Wall Structures connected with dissemination of national knowledge and experiences concerning building masonry enclosure systems.

In Poland, quite similar than in other countries, the building technology in such period is strongly connected with national and regional tradition, level of building industry development and economic growth and specific problems characteristic for the given country. The being in usage masonry wall system is linked to a construction methods and practice tradition connected with changing of building materials industry and introduction of new design rules and requirements. This process is very clearly to observe in case of Polish masonry building enclosures, especially within the period of the last twenty years.

In Poland the one-family houses usually were built of masonry wall structures made of solid clay bricks or some lightweight concrete hollow blocks – very often prepared themselves by investors. Quite different situation was observed in case of multi-family buildings. Up to late 70-ties as typical, the large panel building systems were the dominant. All was changed during last twenty years. It is connected with two factors: political and technical. The political factor is connected with Polish transformation from socialist and economic centralized system to democracy and free market. As the result of these changes the very intensive development of the masonry material industry was observed. Now, the new multi-family buildings, usually the apartment houses are erected as wall-frame structures with concrete frame or slab-column load-bearing structures with internal load-bearing or non-bearing walls and external curtain walls.

In this report, the author presents briefly the analysis of the evolution of masonry walls and building systems occurred on masonry practices with wider describing of some specific problems that are representative in case of Polish, important and interested from research point of view.

2   BUILDING SECTOR AND MASONRY PRODUCTION IN POLAND

2.1   *Importance of the building sector*

Poland is a country in the Middle East Europe. With almost 38 million inhabitants, Poland is the greatest country in that part of Europe. Moreover, since 1989 Poland is a country under transformation process from a socialist central controlled economy to a free-market economy. The construction activity is very important to the economy and represents between 5 and 8% of the Gross Domestic Product (Figure 1) and about 5 to 6% of total employment (see Figure 2).

In Poland there is a high rate of unemployment, between 17 to 18%, in last 5 years. Of course, this is the official level of unemployment. Fortunately, the real situation is a little better. The result of such situation and high cost of work (components of salary connected with social benefits, health, insurance and pension

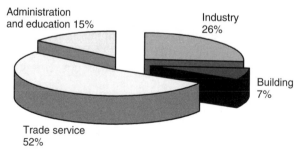

Figure 1.   Overview of economical activities in Poland in 2004 according to Polish statistic data [1].

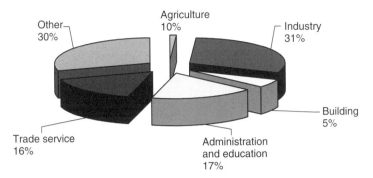

Figure 2.   The employment structure in Poland in 2004 according to Polish statistic data [1].

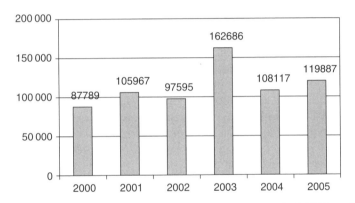

Figure 3.   Number of new dwellings finished and sold in period 2000 to 2005 [2].

premium) is a wide black construction employment, being outside control. Therefore, the real construction activity can be stated much higher than 6% of total employment.

As result of the 50 years of central controlled economy there is a lack of about 3 million dwellings. The large number of Polish citizens is still waiting for their own dwelling, flat, apartment or small one-family house. Figure 3 shows the number of new dwellings finished and sold in period 2000 to 2005, whereas in Table 1 the division of total number of dwellings on types of destiny of utilisation is presented.

According to main focuses of national strategy, especially after Poland joined into EC Members, the problem of insufficient number of dwellings is one of the most important. Therefore, the construction activity in Poland is characterised by a strong development during last 10 years. This sector of Polish economy is expected to have a very quickly growing up for the next 20, or even more, years.

Table 1. Number of new dwellings finished and sold in Poland in period 2000 to 2005 with division on types of destiny of utilisation.

	2000	2001	2002	2003	2004
Co-operative dwellings	24391	25835	15406	11957	9432
Individual dwellings	35542	40642	52433	118034	64858
Dwellings for sale or to let	20728	29403	21970	23844	24230
Others	7128	10087	7786	8851	9597
Total	87789	105967	97595	162686	108117

2.2 *Systems for housing and commerce until 1990s*

a) Systems used before World War II

The construction systems used in terrains contained today's Polish territory were quite similar to other European countries. That situation is strongly connected with Polish history.

At the end of XVIII century, three occupants divided Poland: Prussia, Austria-Hungary and Russia. This situation in case of large part of today's Polish territory has gone on up to 1918. Therefore, now in Poland, we can observe some differences between traditional architecture in north, west and south areas (architecture quite similar to German and Austrian tradition) and in east part of the country, where was significant the Russian influence. Of course, there are quite good visible differences between urban regions (city centres) and rural areas – especially in east part of Poland, which have been always more poor than all terrains with German and Austrian influences (north, west and south Poland).

Generally, in case of traditional housing, commerce buildings and offices in urban regions buildings were usually built of solid masonry walls with timber floor and roof structures (Figure 3). As the dominant was the construction system based on the walls load-bearing solutions. Only monuments or administrative buildings had external plastering and architectural details (see Figure 4a).

Up to 1940s, mainly were used solid clay bricks and lime or cement-lime mortars. Small one or maximum two storey one-family buildings (see Figure 4b) had masonry solid walls, which thickness not exceed 38 to 52 cm – of course, without any thermal insulation. Floor and roof constructions were built of timber elements. The roof surface was usually covered by ceramic roofing-tiles. In case of multi-family housings (practically up to 5 storeys – like shown in Figure 4c and Figure 4d) or other type of buildings (offices, commerce buildings etc.) the wall thickness was greater – even about 95 cm.

b) Systems used after World War II – up to 1980s

After World War II the situation in Polish architecture was specific. With the war, most of city centres were completely destroyed and demolished. Also large number of buildings of the countryside were destroyed or damaged. Therefore, during the first 10 to 15 years after the war end most of building activities were connected with restoration and repair works. In that period as much as possible were used the materials (mainly clay bricks, timber beams and boards, steel elements, etc.) taken from the demolition of damaged buildings and structures.

Figure 4.   Examples of traditional buildings erected before World War II.

Figure 5.   Examples of prefabricated RC large-panel buildings.

From late 1950s started in Poland a new period of construction activity. During the following 30 years in reference to housing the dominant was introduced and developed large-panel building systems. Most of housing was built with that system up to late 1980s. Typically, these systems are multi storey buildings (up to 24 storeys) built of prefabricated enclosures and load-bearing panels (Figure 5).

Figure 6.   Examples of one-family housing built from 1950 to late 1980s.

All load-bearing internal walls were made of solid reinforced concrete panels connected to each other along their vertical edges. These panels were monolithically fixed with prefabricated RC floor plates.

Generally, the thickness of internal walls was from 10 to 15 cm. According to thermal terms each external enclosure panel consist of three layers:

- internal load-bearing layer with thickness of 10 to 15 cm (made of RC);
- layer of thermal insulation with 5 to 12 cm thickness;
- external façade layer – up to 5 cm thickness (also made of RC).

The quality of all prefabricates was usually very poor. The thermal insulation was also not sufficient for good protection of the building interior.

In this period, simultaneously was excising also traditional type of building engineering. Small one-family houses (one or maximum two storeys buildings – see examples shown in Figure 6), were built of masonry enclosure walls and RC floors. This type of construction system of new buildings was popular and widely used.

Unfortunately, during socialist period in Poland, having own house was treated as luxury. People usually were not able to spend enough money for their new house. Therefore most of such buildings erected in this period were built as cheap as possible, with usage of cheap (and bed quality) materials. Also the architectural form was very simple: most of those buildings have a clear cubic form – good from the cost point of view, but very ugly (see Figure 6).

Very often roof construction was very simple. Typically is a solid hip roof consisting of load-bearing floor plate (usually made of RC) without any thermal insulation (see icicle shown in Figure 6) with surface covered by steel sheets or bitumen roofing paper.

Masonry walls were usually made of solid clay bricks or lightweight concrete hollow blocks (self made by the house owner) and also AAC blocks designed to use common joints made of cement-lime mortar.

The quality of building materials, especially AAC blocks and all self-made concrete hollow blocks was usually very poor and characterised by low mechanical properties. External walls consisting of single-layer structures were offered sufficient resistance but not sufficient thermal insulation. Finishing of the external wall surface was made of cement mortar, sometimes painted.

2.3   *Modern Polish building systems*

The Polish construction activity was completely changed from the beginning of 1990s, when the socialist period of our history was finished. Since that time the extremely intensive development of building market is observed. This situation is connected with two factors:

1. introducing into the Polish building market new building materials and technologies;
2. green light for establishing of private building companies.

As mentioned, in Poland there is a lack of about 3 million dwellings. Therefore building construction has been the most important construction activity and, within this group, the housing buildings are the dominant. The second very important sector of construction activity is the construction of bridges and roads, specially, highways.

Figure 7 shows the structure of enclosure walls buildings 65% of total made new walls belong to housing. Quite similar situation is observed in case of floor and roof construction (Figure 8). That means that housing building is now the most important and fast developed sector of construction activity in Poland.

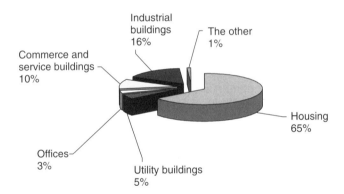

Figure 7.   The structure of enclosure walls building [3].

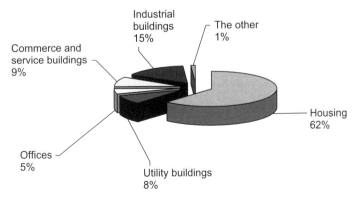

Figure 8.   The structure of floor and roof construction [3].

Figure 9.   Examples of modern multi-storey buildings with RC load-bearing frame construction and masonry infill.

Multi-storey buildings (housings, offices or commerce buildings) nowadays in Poland are usually constructed as RC frame or slab-column structures with masonry infill (Figure 9). Sometimes the frame load-bearing structures are made of steel. This is quite new situation.

Masonry walls used as infill of frame or slab-column in major part are typically non-load walls which main role is to protect the interior against external environment. Therefore, now, the structure of such walls should be adequate to that new function. In opposite to traditionally masonry buildings (erected up to half of XX century), in case of enclosure walls which the most important function is thermal and acoustic insulation, other than good resistance, properties required are: lightness, good thermal and sound insulation, easiness of use and, finally, low cost.

Taken into consideration all specified above characteristics, the very intensive development of masonry unit's market in Poland since late 1980s is absolutely comprehensible.

Nowadays, the non-load external walls in Poland are built of hollow blocks (clay or calcium silicate), or AAC blocks with thin joints. There are usually single leaf walls with external thermal insulation covered by stucco, or three-leaf cavity walls with internal leaf built of hollow or AAC blocks and external leaf usually made of solid clay or calcium silicate bricks, without plastering. The total thickness (including all leafs) of such walls is between 25 to ca. 50 cm.

Sometimes, in case of buildings with RC or steel load-bearing construction is necessary to ensure the suitable space stiffness. In such cases some enclosure and internal walls must be load-bearing and made of masonry units characterised by higher mechanical characteristics and with usage of higher class cement-lime mortars. In Poland the lowest mortar class for load-bearing walls is M5 (mean compressive strength of $5\,N/mm^2$). Thickness of that type walls is usually 19 to 45 cm.

The large part of nowadays construction activity in Poland is connected with one-family with one or two above-ground storeys. Also in this type of buildings the new architectural and construction solutions are observed. Generally, the major part of small (with usable area between 90 and $200\,m^2$) one-family housing is constructed with wall load-bearing system. Enclosure walls are usually built as single leaf, usually single leaf with external thermal insulation covered by stucco, or three-leaf cavity walls.

(a)

(b)

Figure 10. Examples of modern one-family housing erected in Poland: (a) free-standing housings (under construction); (b) row housings.

The most popular are different types of clay hollow blocks or modern AAC blocks with thin joints (Figure 10). The tradition of load-bearing walls or part of walls construction with solid clay bricks was replaced by solutions based on hollow blocks. Much less used are calcium silicate hollow blocks and all types of lightweight hollow concrete blocks. Mostly, these types of masonry units are in use in east and east-south Poland.

Floors in small one-family housing are usually constructed as RC structures. There are many types (also with prefabricated beams and floor blocks) of such floors in use in Poland. One of the most popular (and cheap) is simple RC solid floor plate.

Modern housing, built since late 1980s have mainly timber roof construction with adequate thermal insulation and roofing made of different types of ceramic, concrete or steel roofing tiles (see examples of houses shown in Figure 10).

3 MOST COMMON ENCLOSURE WALL SYSTEMS

3.1 *Materials for enclosure masonry walls*

a) Masonry units

In Polish building material market a lot of types of masonry units are available. Among big building material companies (operating in whole Europe or even over

Table 2.   Examples of most popular masonry units in Poland.

Type and shapes	Dimensions (length × height × width) (cm)	Material
Solid bricks		
	25 × 12 × 6.5	Clay Concrete Calcium silicate
Hollow blocks		

	19 × 19 × 24
	24 × 24 × 29
	34 × 24 × 19
	44 × 24.7 × 23.8
	36.5 × 24.7 × 23.8
	30 × 24.7 × 23.8
	24 × 37.2 × 23.8
		Clay
		Lightweight
		  concrete
		Calcium
		  silicate

AAC blocks		
	19 × 19 × 59 24 × 24 × 36.5 24 × 24 × 59	Autoclaved aerated concrete

the ocean) there are still large number of small factories and companies, like e.g. brickyards. All these small building enterprises, now quite private, produce traditional masonry units in form of solid bricks or hollow blocks known and already used in building practice for many years. The quality level of production in still very often rather low, but is improving every year.

Generally, the most popular materials in Polish building market are (see also Table 2):

– clay units, solid or vertically perforated, usually used on enclosure and internal load-bearing walls (classified as group 1 and group 2, according to EC6);
– AAC solid blocks used in enclosure walls and also as load-bearing internal walls in small up to two above-ground storeys buildings;

– dense or lightweight concrete vertically perforated units, used in enclosure and internal load-bearing walls (classified as group 2, according to EC6);
– calcium silicate solid and vertically perforated units, not yet so popular, but sometimes used in outer leaf of cavity walls or internal load-bearing walls (classified as group 1 and group 2 according to EC6).

Practically, every year are introduced into Polish building material's market new kinds of masonry units. A very strong development on that field is observed.

b) Mortars
Formerly in Poland, mortars were usually made by workers, on site, as Portland cement: lime:sand mixes. Two types of lime were used: hydrated and hydraulic. With those components have been preparing typically general purpose mortars.

Mixes are, generally, not rich. The most popular and wide used is mortar 1:1:6 (Portland cement:hydrated or hydraulic lime:sand, by volume). Sometimes, when high masonry strength is necessary, more rich mortar mix, usually, 1:3 (Portland cement:sand), is used.

During last 10 years situation is changing. More and more popular are usage of ready manufactured mixes of suitable strength classes (e.g. M5, M7 or M10).

c) Ancillary components
In case of cavity walls is necessary using of ties to fix to each other both leafs of the wall. Former in Poland in such situations some problems are appeared. Simply, in building market was very difficult to find ties made of steel protected against corrosion.

Last time the situation was distinctly improved. In building market were introduced modern systems of ties (see Figure 11a) made of galvanised or stainless steel, compatible with European standard EN 845-3:2002 [6] requirements.

Since 1998 in Poland is also available a type of special prefabricated reinforcement to be used in bed joints, also in thin bed joints. This is truss type reinforcement (see Figure 11b) made of galvanised or stainless steel.

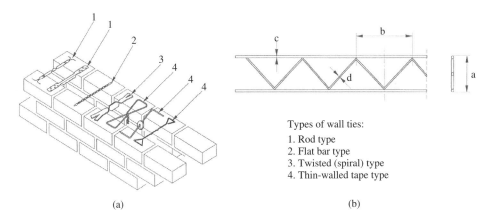

Types of wall ties:
1. Rod type
2. Flat bar type
3. Twisted (spiral) type
4. Thin-walled tape type

(a)   (b)

Figure 11.   Ancillary components used in masonry enclosure walls: (a) wall ties; (b) prefabricated bed joint reinforcement.

3.2   *Most common masonry building enclosures*

As referred, Poland is situated in Central Europe. The climatic circumstances are typically of continental areas. Summers are rather dry with maximum temperatures in July and August often exceeding 30°C and winters are very often cold with sharply falling minimum temperatures in December and January below –20°C. Additionally, very often winters are very snowy and windy. Practically, this type of climate is representative for whole territory of Poland excluding the parts along to seashore.

The result of such climatic circumstances is requirement of good thermal insulation of all external walls. There are two ways to fulfil this requirement:

1. increasing of wall thickness;
2. usage of external thermal insulation.

The increasing of wall thickness is still popular in Poland. By comparison with building tradition during XX century, introducing into building material's market many new types of hollow blocks allowed to construct walls as single leaf with quite sufficient thermal insulation. Unfortunately, the thickness of such walls is 38 to 50 cm and do not eliminate the thermal bridges existing, e.g. along the girder beams. Therefore, the thermal comfort inside building is not so satisfying.

The second solution (see Figure 12) is much better. Usage thermal insulation as a component of enclosure external wall permitted to protect interior of the building against all weather influences and additionally make possible to eliminate most of thermal bridges. Therefore, this type of walls is now the most popular and willingly used.

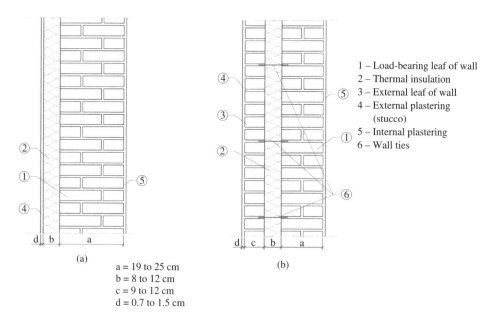

1 – Load-bearing leaf of wall
2 – Thermal insulation
3 – External leaf of wall
4 – External plastering
  (stucco)
5 – Internal plastering
6 – Wall ties

a = 19 to 25 cm
b = 8 to 12 cm
c = 9 to 12 cm
d = 0.7 to 1.5 cm

Figure 12.   Typical cross-section of modern enclosure wall: (a) one leaf wall; (b) three-leaf wall.

The external walls are usually built of hollow blocks (clay or calcium silicate) or AAC blocks with thin joints. Two construction solutions are being in use:

1. Single leaf walls with external thermal insulation covered (see Figure 12a) by stucco.
2. Three-leaf cavity walls (shown in Figure 12b) with cavity completely filled with thermal insulation and both wall leafs in masonry.

In case of one-family housing, the typical thickness of load-bearing external single walls (without thermal insulation) made of clay bricks are between 38 and 52 cm. When the additional thermal insulation is done, from resistance point of view thickness of this part of wall can be reduced to 25 cm. Situation is quite different, when walls are built of different types of hollow blocks. In these cases, very often wall is thick (37 to 45 cm) and made without any thermal insulation.

The typically total thickness (including all leafs) of cavity walls is between 30 to ca. 50 cm. The internal leaf of cavity walls is often built of hollow clay or AAC blocks whereas the external one is usually made of solid clay bricks or calcium silicate blocks, without plastering.

4   EVOLUTION TRENDS

4.1   *Repair and thermal renovation*

In Poland, the requirements connected with thermal protection of the buildings until 1980s were ca. three times lower that being in force today. Generally, all enclosure walls, floor and roof structures should be constructed in form which has guaranteed not reflux appearance of the wall or floor surface.

Situation is completely changed beginning since 1982, when the first national standard concerning thermal protection of buildings PN-82/B-02020 [7] was elaborated and established. At the first time the special detailed requirements connected with thermal resistance of enclosure walls, floors and roof constructions were established. As the result of these new regulations, the necessity of thermal insulation materials using was appeared. In 1991 standard [7] was modified. The thermal conditions for new erected and restored buildings have been sharpening.

In Figure 13 the percentage participation in each type of expenses in the "life cycle of building" is shown. There is important, that energy expenses are more than doubled in comparison with total cost of construction process. Moreover, the currently utilization expenses are comparable with energy or total repair costs. Therefore in Poland since beginning of 1990s the wide process of thermal renovation of buildings are observed. This process concerns both small one-family housings and multi-storey large-panel buildings.

The next step in direction of thermal properties of buildings has done at the end of last century, when the philosophy of thermal protection of buildings was introduced.

In 2001 the new Polish code PNEN 832:2001 [8] was established, which is the Polish national version of the European standard. According to these new regulations not only an adequate thermal resistance of enclosure walls, floors and roof

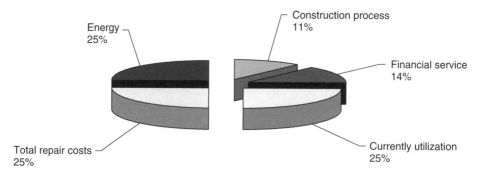

Figure 13.   The percentage participation in each type of expenses in the "life cycle of building" [4].

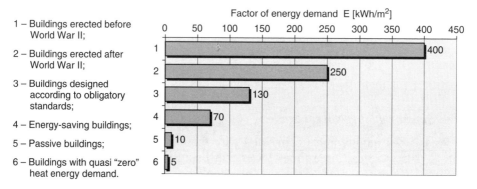

Figure 14.   The mean values of "seasonal heat demand factors" E for different types of buildings [5].

constructions are important but the main became the "seasonal heat demand factor" E (see Figure 14), where the mean values of E for different types of buildings are shown. Practically, all buildings erected before 1990s need thermal renovation. Therefore, large part of building activity and expenses are just connected with this type of action.

The secondary positive effect of thermal renovation of existing buildings, specially erected between 1950 and 1990, are usually the significant improvement of architectural looks. In Figure 15 the comparison of appearance of ca. 50 years old typically one-family building before and after thorough thermal renovation is presented.

4.2   *Natural seismicity*

Generally is accepted that Poland is a non seismic country. Therefore, there is no seismic code or other special requirements for protection of new erected or modernised buildings against this type of dynamic influences.

(a)   (b)

Figure 15.   Example of 50 years old typically one-family building before (a) and after (b) thorough thermal renovation.

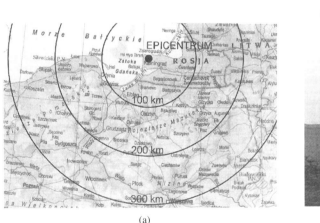

(a)   (b)

Figure 16.   Earthquake of 21st September 2004 observed in North-East Poland: (a) the range of tremors; (b) example of damaged building.

Unfortunately, territory of Poland is running a risk of earthquakes. The last two most intensive earthquakes were observed in Poland at the end of 2004. The first one took place in North-East Poland and was characterised by two main earth tremors (with magnitudes of 4,8 and 4,9 according to Richter scale [9]). The epi-centre of this earthquake was located in city Kaliningrad (Russian territory between Poland and Lithuania), see Figure 16a. As the result of this phenomenon damages of more that 100 buildings on Polish territory were recorded. Some of them were quite serious, see Figure 16b.

The second earthquake, on 30th November took place in South Poland with epicentre close to Slovak Republic border. The magnitude of this phenomenon was estimated as 4,6 according to Richter scale [10]. The damages of buildings, mainly built as typically small masonry constructions with wall load-bearing system were observed in many mountain resorts.

Besides powerful earthquakes, like described above from 2004, every year in Poland more than hundred weak earth tremors are recorded. This type of dynamic influences usually took place in South-West Poland. The magnitude of such tremors is not so high (not exceeding 3 in Richter scale).

Nowadays, the situation is not changed. There are still lack of suitable standards and requirements related with protection of new erected buildings against earthquakes. Probably, only introducing into building design practice the national version of Eurocode 8 will improve the situation in this domain.

4.3   *Specificity of building on terrains subjected to mining influences*

a)   Description of the problem

The problem of building on mining terrains, mainly the buildings with masonry load bearing walls is in Poland considered from many years. The row of recommendations in effect were elaborated in form of instructions [12], [13], [14] widely applied in design practice. These studies concern the methods of buildings design on mining terrains mainly restricted to correct fulfilment of limit states for foundations and underground parts of the buildings. Based on these instructions is not possible the accomplishment of full strain and stress analysis of stiffening masonry walls of the building during design process.

Only the Polish Code PN-B-03002:1999 [15] gives the method of limit state checking for masonry walls subjected to irregular vertical ground displacements based on analysis of non-dilatational strain angles. This method can be applied also in reference to buildings erected on mining terrains.

b)   Mining influences as loads acting on the building

One of main influences which should be considered in analysis of construction, according to Eurocode 7 [11], are displacements caused with mining exploitation. The underground exploitation of minerals is connected with deformation and displacements of the orogenic belt which is the factor of the terrain deformation appearance.

From the buildings protection point of view, the largest problem is the non-continuous deformations and tremors because they have the character of difficult to earlier foreseeing random events, and results of their influence on the constructions is often connected with large damages.

Some different matter looks in addition to continuous ground deformations, the most universally stepping out, which parameters can be estimated. Instruction No 12 [13] published by Main Institute of Mining (GIG) and then Instruction No 286 [12] edited by Building Research Institute (ITB) were given 5 mining categories (from I to V), meanwhile in instruction ITB No 364/2000 [14] one category more was added, appointed as "0" – with the lowest values of parameters of deformation, what is shown in Table 3.

In the analysis of load-bearing walls of masonry buildings subjected to irregular vertical ground displacements can be accepted mainly only one coefficient: the radius

Table 3.   Categories of deformation of mining terrain according to [14].

Category	Tilt T (mm/m)	Radius of curvature R (km)	Horizontal strain ε (mm/m)
	Ultimate values of deformation parameters		
0	$T \leqslant 0.5$	$\lvert R \rvert \geqslant 40$	$\lvert \varepsilon \rvert \leqslant 0.3$
I	$0.5 < T \leqslant 2.5$	$40 > \lvert R \rvert \geqslant 20$	$0.3 < \lvert \varepsilon \rvert \leqslant 1.5$
II	$2.5 < T \leqslant 5$	$20 > \lvert R \rvert \geqslant 12$	$1.5 < \lvert \varepsilon \rvert \leqslant 3$
III	$5 < T \leqslant 10$	$12 > \lvert R \rvert \geqslant 6$	$3 < \lvert \varepsilon \rvert \leqslant 6$
IV	$10 < T \leqslant 15$	$6 > \lvert R \rvert \geqslant 4$	$6 < \lvert \varepsilon \rvert \leqslant 9$
V	$T > 15$	$\lvert R \rvert < 4$	$\lvert \varepsilon \rvert > 9$

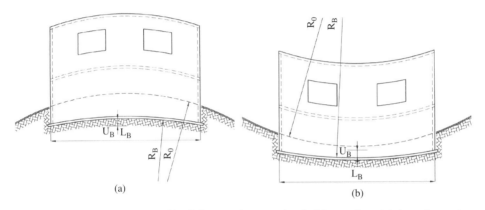

Figure 17.   Deformation of building and ground subsidence for: (a) hogging mine basin; (b) sagging mine basin.

of curvature R. For not so tall masonry building (usually up to 4 above-ground storeys) two other parameters, i.e. the tilt of the terrain T and horizontal strain ε, have smaller influence.

The internal forces produced by horizontal strains ε should be taking over by suitable protection of foundations and didn't give a vertical component of deformations. The excessive tilt of building can produce into load bearing walls the additional component of internal forces – and in consequence – the larger local ground pressures. Situation like that can be connected with appearance of additional vertical component of settlements. Usually they are considerably smaller than values of settlements connected with the curvature of terrain K that can be characterized by radius of curvature R. There are two types of curvature:

(a) convex curvature – Figure 17a;
(b) concave curvature – Figure 17b.

The radius of bottom surface of underground storey flexion (the foundations) R_B is usually considerably larger than R_0 (see Figure 17). It is connected with building-subsoil interaction.

Usually, in design practice the curvature described by radius R_0 is taken as a correct quantity. It is accepted, that this radius answers the average shape of deformed surface of the terrain. In reality can happen, that on the length of considered building (or the given load bearing wall) the local values of curvature can be different. That means that the shape of deformed surface of the subsoil in some areas can not agree with curve described by the radius R_0. The envelope in exact solution was received with gathering of local curvatures.

c) Analysis of masonry stiffening walls – the strain criterion

According to regulations given in instruction [12] the load state created by coal mining ground deformations is taken as short-time variable loads. The *basic combination of loads*, consisting from uniform and variable loads is obligatory in design of buildings subjected to influences of continuous subsoil deformations. Whereas in analysis of buildings running a risk of non-continuous deformations it should be taken into account *the exceptional load combination*.

On the basis of many years of investigations carried out at Silesian University of Technology to the new Polish Masonry Code PN-B-03002:1999 [15] the strain criterion permitting to analyse the state of strain and stress of walls subjected to vertical subsoil displacements through checking the serviceability state condition was introduced. It specifies the methodology of limit state checking of masonry load bearing (stiffening) wall on the basis of their state of deformation.

In masonry stiffening walls, with attention on irregular, vertical subsoil displacements (including produced by mining influences) the non-dilatational strains and the concurrent to them main stresses are the predominant. As the result of such state of stress and strain the diagonal cracks appear on the masonry wall.

The level of vertical stresses has the essential influence on the values of principal tensile stresses and diagonal cracks appearance. It can be taken that the influence of wall deformations, created by displacements difference of it's both edges, on compressive stresses connected with vertical loads is very small. It mainly depends on values of vertical loads (self weight of the wall and all loads transmitted to calculated wall from upper floors), acting in the analysed wall. In connection with given above on specific of calculation of walls subjected to this type of the vertical displacement decide the serviceability limit state.

The condition of not exceeding the serviceability limit state (according to [15]) is as follow:

$$\Theta_{Sd} \leq \Theta_{adm} \tag{1}$$

where Θ_{Sd} – non-dilatational strain angle (determined for characteristic values of loads); Θ_{adm} – admissible value of non-dilatational strain angle.

In case of walls in building subjected to irregular subsoil deformations (including buildings situated on coal mining terrains), the Θ_{Sd} can be calculated in approximate way on the basis of geometry of deformed wall. Design scheme is shown in Figure 18.

In such situation the following formula may be used:

$$\Theta_{Sd} = \frac{\Delta a}{L}$$

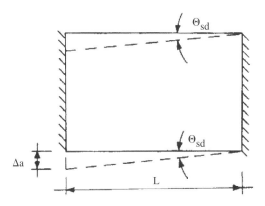

Figure 18.   Scheme for Θ_{Sd} determining in case of stiffening walls subjecting to irregular subsoil deformations according to [15].

Table 4.   Admissible values of non-dilatational strain angles according to [15].

Type of masonry unit	Cement mortar	Cement-lime mortar
Group 1 except AAC blocks	0.4 mm/m	0.5 mm/m
Group 2 and Group 3	1	1
AAC blocks with common joints	0.2 mm/m	0.3 mm/m

[1] Values should be determined by tests.

where Δa – displacements difference of two transverse load bearing walls; L – distance between transverse walls.

Determination of displacement difference, Δa, is usually not so simple, because these displacements are function of the stiffness of whole load bearing system of the building.

d)   The admissible values of non-dilatational strain angle
The obligatory Polish Masonry Code [15] gives (see Table 4) in suitable table admissible values of non-dilatational strain angles Θ_{adm}. These values were taken mainly on the basis of investigations carried out at Silesian University of Technology [16] and concern the situation, when as a criterion of not exceeding the serviceability limit state is taken that the masonry wall subjected to influences produced vertical shearing forces stays still unreinforced or width of appeared cracks does not exceed 0.1–0.3 mm.

This is too "sharp" condition. In reference to buildings running a risk of influences connected with mining activity that condition is practically not possible to satisfy. In case of buildings was introduced [19] the *"temporary serviceability limit states"*, i.e. in specified period is allowed appearance of cracks which width exceeds values given above. Therefore, by the author's sentence [17], [18], for buildings situated on mining terrains, in situation when can appear, in limited time and the range, deformations which can produced small damages would be correct to allow temporarily higher Θ_{adm} values than given in [15], in case of walls made

Table 5. Comparison of Θ_{adm} values calculated using (2) for masonry made of group 1 masonry units and cement-lime mortar with values taken in [19] and [14].

	Effects in building		Θ_{adm} formula (2) (mm/m)
	According to [19] and instruction [14]		
Stage of difficulty	Θ_{adm} (mm/m)	a_w (mm)	
Not noticeable	<1.0	<1.0	<1.08
Small	1.0 ÷ 2.0	1.0 ÷ 5.0	1.08 ÷ 1.84
Medium	2.0 ÷ 3.0	5.0 ÷ 10.0	1.84 ÷ 2.32
High	3.0 ÷ 5.0	10.0 ÷ 30.0	2.32 ÷ 3.35
Excluding use	>5.0	>30.0	>3.35

of masonry units of group 1 with cement mortar or cement-lime mortar, calculated by the following formula:

$$\Theta_{Sd} \leq \sqrt[3]{b_T \frac{a_w}{0,3}} \cdot \Theta_{adm} \tag{2}$$

where Θ_{Sd} – non-dilatational strain angle determined by geometrically analysis of building subjected to vertical mining displacements; Θ_{adm} – admissible value of non-dilatational strain angle (taken from Code [15]); a_w – temporary permitted value of cracks width (in millimetres); b_T – coefficient taking into account the speed of mining exploitation; $b_T = 1$, in case of quick exploitation (mining basin progress over 20 m per twenty-four) or in analysis of buildings subjected to influences connected with non-continuous deformations; $b_T = 3$, in remaining cases.

Putting into formula (2) $b_T = 3$ as well as the values of admissible crack widths, given in work [19], for individual stages of difficulties is received (for walls from the units group 1 with cement-lime mortar) the values of non-dilatational strain angles quite similar like taken in instruction [14], what it was shown in Table 5.

From authors experiences come that load-bearing walls of buildings subjected to vertical shearing should not be preparing with usage of masonry units of group 3 without regard on class and the mortar type, at least to time, when the suitable experimental investigations of the behaviour of this type of walls as well the safe values of admissible non-dilatational strain angles will be determined.

5 REFERENCES

[1] Central Statistical Office. Bulletin No 12, January 2006.
[2] Central Statistical Office. Construction – Activity Results in 2004.

[3] Sochacki, M. – "Buildings of walls and floors". *Materialy budowlane*, No 4/2002 (in Polish).

[4] Lange, I. – "The way to energy-saving – Intelligent buildings". *Przeglad budowlany*, No 2/2002, (in Polish).

[5] Dubas, W. – "Basis of energy-saving buildings". *Przeglad budowlany*, No 10/2005 (in Polish).

[6] EN 845-3:2002: "Ancillary components".

[7] PN-82/B-02020: "Thermal protection of buildings. Requirements and calculations".

[8] PNEN 832:2001: "Thermal utility properties of buildings – Calculation of heat energy demand – Housings".

[9] Zembaty, Z.; Jankowski, R.; Cholewicki, A.; Szulc, J. – "On two earthquakes affecting the North-East Poland and their influence on construction objects". *Inzynieria i Budownictwo*, No 1/2005, pp. 3–9 (in Polish).

[10] Zembaty, Z.; Jankowski, R.; Cholewicki, A.; Szulc, J. – "On earthquake on 30th November 2004 in the region of Podhale, and its influence on building structures". *Inzynieria i Budownictwo*, No 9/2005, pp. 507–511 (in Polish).

[11] ENV 1997-1:1994 – "Eurocode 7. Geotechnical Design. Part 1: General rules".

[12] Instruction No 286 – "Design directives for buildings with wall load-bearing system subjected to mining exploitation influences". Warsaw, 1989 (Annex 1993) (in Polish).

[13] Instruction No 325 – "Design of housings and public buildings subjected to mining tremors influences", ITB, Warsaw 1993 (in Polish).

[14] Instruction No 364/2000 – "Technical requirements for building objects situated on mining terrains", ITB, Warsaw 2000 (in Polish).

[15] PN-B-03002:1999 (with Az1/2000 and Az2/2002) – "Unreinforced masonry structures. Analysis and structural design".

[16] Kubica, J. – "Unreinforced masonry walls subjected to non-dilatational strains produced by irregular vertical ground displacements". Politechnika Slaska Publishing, Gliwice, 2003 (in Polish).

[17] Kubica, J. – "The Polish Approach to the Shear Walls Analysis". Proc. of the 35th Meeting of *CIB/W02 – Wall Structures Commission*, Dresden, October 1998, CIB Publication No 242, pp. 89–98.

[18] Kubica, J. – "Walls of masonry buildings subjected to irregular settlements. Proposition of a supplement to Eurocode 6". Proc. of the *13th IBMaC*, Amsterdam 2004, pp. 1215–1224.

[19] Kawulok, M. – "Evaluation of serviceability properties of buildings considering mining influences". ITB Publishing, Series: Transactions, Warsaw 2000 (in Polish).

CHAPTER 11

Typical masonry wall enclosures in Portugal

Hipólito de
SOUSA

*Associate
Professor FEUP
Porto
Portugal*

Fernanda
CARVALHO

*Senior Research
Officer LNEC
Lisboa
Portugal*

SUMMARY

Practices and problems related with masonry building enclosures are presented. After a short review of recent Portuguese evolution concerning housing buildings, a description of the most frequent structures, masonry materials and enclosures is presented and illustrated. The main problems and pathology related with these practices and the evolution trends are summarized.

1   INTRODUCTION

According to the strategy defined on CIB Commission W023-Wall Structures, an effort to disseminate national perspectives related with building masonry enclosure systems has been considered of interest.

Masonry is perhaps the building technology more deeply affected by regional and traditional practices concerning materials, detailing and construction. In order to understand the masonry specialists' concerns in different countries, it is fundamental to know the specificities and real problems of these subjects in each country.

The purpose of this report is to present briefly the recent Portuguese situation concerning this matter. The paper shows that in the last few years a quick evolution occurred on masonry practices, which produced several changes and some problems that must be well studied and explained, opening therefore important and interesting research perspectives.

2   BUILDING SECTOR

2.1   *Importance of the building sector*

Portugal is a South European country, Mediterranean on the Centre and South, but with increasing Atlantic influences on the Northwest. The population, of almost 10 million inhabitants, is concentrated near the sea. The construction activity is important to the economy and represents about 7% of the Gross Domestic Product and 9% of employment [1]. Until the end of the 1990s, buildings were the most important construction activity; this effort being focused on new buildings, Figure 1[2]. Nevertheless, at the moment this effort is decreasing.

2.2   *Structural systems for buildings*

After the World War II, building solutions and technologies have quickly evolved, the traditional practices being progressively replaced by new ones, not always adapted to local conditions as formerly.

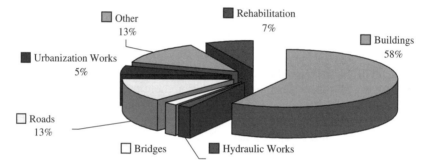

Figure 1.   Importance of construction sectors according to 1998 Portuguese statistic data [2].

Figure 2.   Example of typical buildings with reinforced concrete framed structure.

Concerning existing structural systems the major building structures are of reinforced concrete frame (Figure 2). The other solutions, like steel, masonry and timber structures are seldom used, even in small and one family houses. Also the structures with reinforced concrete walls are only used in a few special buildings, or where the fixing of heavy claddings imposes its use.

Some minor attempts to change this situation are under way. In Civil Engineering and Architecture Schools the attention paid to alternative structural solutions (steel, masonry and timber) is growing, supported by the Eurocodes and international images and models. Furthermore, producers and sellers see in these alternatives and new materials good business opportunities.

2.3   *Systems used for housing and commerce*

As mentioned, in the past the traditional Portuguese housing architecture usually presented regional solutions, well adapted to climatic conditions. The use of stone in heavy and thick walls was predominant.

Except in some rustic regions or in more primitive buildings of the countryside, where the stone remained not finished, usually the stone enclosure walls were covered by a thick porous render, with a low modulus of elasticity and made in multiple layers by very skilful workers.

In the Atlantic coast, in the zones more exposed to rain, the watertightness of the render was usually improved by the introduction of a waterproof layer, directly applied on the support, made of asphaltic mortar, or of a very rich Portland cement mortar with a hydrofuge admixture. Ceramic decorative coverings and claddings made of slate tiles or asbestos-cement profiled sheets, mainly on gables, were used (Figure 3) [3], with the same purpose of improving the watertightness, but with a more regional character.

Figure 3.   Examples of Portuguese traditional buildings.

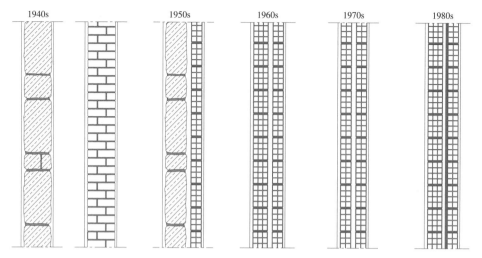

Figure 4.   Synthesis of the evolution process of Portuguese walls [3].

By the end of the 1940s, and mainly in urban regions, the use of concrete structures would become widespread, first on the floors replacing timber, being progressively extended to the vertical support elements. At the time, the walls lost their resistance role and became only infilling elements, the stone being replaced by clay bricks. Clay brick producers developed a multitude of shapes, but always progressing from solid to large horizontally perforated elements.

The importance of raintightness, associated to the thickness and weight wall reduction, led in the 1960s to the generalization of enclosure cavity walls made of clay bricks, introduced with adaptations from abroad. In the 1980s, the concern for thermal comfort and energy conservation and the consequent publication of the respective code [8], led to the vulgar use of thermal insulation products filling the air space of cavity walls.

In this evolving process, the tradition of rendering the walls with cimentitious mortars remained, but the quality of execution has decreased and the regional character of architectonic solutions has been lost (Figure 4). Although in this evolution the masonry walls have lost their structural importance, they remain an important construction element both in functional and economic terms.

Concerning other buildings, like commercial, industrial and service ones, the solutions are generally the same, but the embodiment of steel and concrete precast concrete structures is higher.

The partition walls were traditionally made using thin wooden elements plastered with hydrated lime mortar, but have been progressively replaced by thin clay brick walls, with the thickness of the bricks being referred to on Figure 5.

The importance of requirements associated with the productivity in the works and lightness of the partitions walls, led to the introduction of other solutions, which are being increasingly used, such as gypsum wallboard, gypsum blocks and other light partition systems.

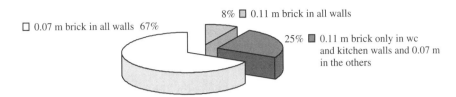

Figure 5. Thickness of clay bricks in current partition walls, (by increasing order) according to a statistic study [4].

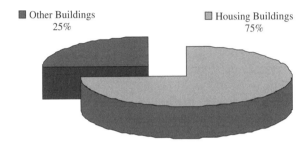

Figure 6. Building construction in Portugal.

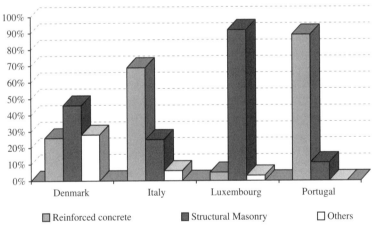

Figure 7. Comparison of building structures in some European countries [2, 5, 6, 7].

2.4 *Some statistics on use and relevance*

As mentioned, building construction has been the most important construction activity and, within this group, the housing buildings represent almost 75% (Figure 6).

Concerning structural systems and according to Portuguese statistic data, the importance of reinforced concrete is close to 90% and is the solution used in the most important and large buildings. Normally, concrete is cast in place. Currently, only some components of the slabs are precast.

Figure 7 shows the comparison of structural systems used in a few European countries. The predominance of concrete structural system in Portugal is obvious.

3   MOST COMMON ENCLOSURE WALL SYSTEMS

3.1   *Masonry materials*

a)   Units

In the Portuguese market, current masonry units (traditional) and some new emergent units (not traditional) are generally available. The industrial level of Portuguese factories of masonry materials is generally good, but there is no sufficient effort towards development, characterization and quality control of the products.

The traditional material producers only commercialise the units and not all the components of masonry system. This leads to the fact that usually the singular points of walls are solved on site by improvisation. The traditional masonry materials are:

– clay units, horizontally perforated, used largely on enclosure and internal walls (group 3 according to EC6);
– clay units, solid or vertically perforated (facing bricks or not) used only in external walls (group 1 and 2 according to EC6);
– aggregate concrete units, dense or lightweight, used more in external walls (group 2 according to EC6);
– natural stone, of which the use is limited to localized regions, in outer leaf of cavity walls.

The most popular materials are the clay units, horizontally perforated, which represent more than 90% of the units used in walls.

The "new" masonry materials, not traditional, are:

– split dense aggregate concrete units;
– lightweight clay units, vertically perforated;
– autoclaved aerated concrete units.

These units, with the exception of split concrete blocks, are not currently produced in Portugal.

Generally, these "new" products are well developed and studied, but their cost is higher than traditional ones. Furthermore, certain conservativeness exists that makes it difficult to accept new materials and solutions. The importance of these products in terms of market is residual.

Tables 1 and 2 show the range values of the significant characteristics of traditional materials.

b)   Mortars

In Portugal, masonry mortars are usually made on site with Portland cement and sand. The use of lime, either hydrated or hydraulic, although frequent in the past, is not common at the moment.

The mixes are generally rich (1:3 or 1:4, cement: sand, by volume). Usually Portland cement type II 32.5 is used. There is no masonry cement on the market. The use of admixtures to improve watertightness is frequent on facing masonry. The use of factory-made mortar (usually dry mortar, supplied into silos) is increasingly fast, mainly in works carried out by medium and large companies.

Table 1. Characteristics of clay units [9, 10, 11].

Dimensions and shapes (length × height × width) (cm)		Weight approx. (kg)	Volume of holes (%)	Compressive strength[2] (MPa)
Horizontally perforated	30 × 20 × 22[1]	7–11	55–70	1.9–3.9
	30 × 20 × 15[1]	5–7	50–65	2.5–4.9
	30 × 20 × 11[1]	4–6	50–65	2.8–5.2
	30 × 20 × 9	3.5–5.5	40–60	3.0–5.7
	30 × 20 × 7[1]	3–5	40–60	3.7–7.0
Vertically perforated	30 × 20 × 4	2–3	40–50	6.0–7.0
	22 × 11 × 7[1]	1.5–2.5	25–40	8.0–9.5
	22 × 11 × 5	1.2–1.7	25–40	8.0–9.5
Solid	22 × 11 × 7[1]	2.5–3.5	–	17.0–48.0

[1] Sizes according to Portuguese standard NP 80.
[2] Expressed in terms of gross area of the specimens, not normalized by shape factors.

Table 2.   Characteristics of concrete blocks [9, 11].

Dimensions and shape (length × height × width) (cm)	Weight approx. (kg)	Volume of holes (%)	Compressive strength[1] (MPa)
(50 or 40) × 20 × 30	20–29	45–65	3.5–4.5
(50 or 40) × 20 × 25	20–25	45–65	3.0–4.5
(50 or 40) × 20 × 20	15–22	40–50	3.0–4.5
(50 or 40) × 20 × 15	12–18	40–50	4–5
(50 or 40) × 20 × 12	12–15	40–50	4–5
(50 or 40) × 20 × 10	10–13	30–50	4–5
(50 or 40) × 20 × 8	8–12	30–50	4–6
(50 or 40) × 20 × 5	8–10	–	6–8

[1] Expressed in terms of gross area of the specimens, not normalized by shape factors.

Figure 8.   Examples of ties and anchors used in Portugal.

c)   Wall ties and reinforcement
The use of wall ties and reinforcements is not current in masonry walls. There is
no national production of this kind of materials for masonry; in a few cases,
imported materials are used. Very often, builders improvise wall ties on site with
galvanized wire.

The ties in cavity walls are in some cases specified by the designer (generally the
architect), but on site they are not placed. Sometimes only some wires of dubious
effectiveness are placed. The use of rigid boards of thermal insulation in cavity walls
increases the difficulty associated with the correct placement of ties (Figure 8).

In lintels, generally some bars are included in the mortar joint. The use of prefabricated reinforced concrete lintels is frequent, in particular for the inclusion of the
roller blind. As the Portuguese masonry walls are generally of simple infilling, the
masonry attachment to the structure should be improved by anchors, which are only
used in a few cases, fig.8.

3.2   *Thermal insulation*

Although the Portuguese climate is moderate and the 1990 thermal code requirements
[8] are not very severe (U value for walls in the coldest region $\leqslant 0.95$ W/(m²°C)[1]),
the awareness of the need for thermal comfort in winter and summer is a reality
today, leading to the generalized use of thermal insulation.

In most common cavity walls thermal insulation is provided by filling totally or
partially the space between the leaves. In single leaf walls, an external insulation
is applied between the finishes and the body of the wall. In some cases, the use of
units with improved thermal behaviour makes it possible to fulfil largely the 1990
thermal code requirements without thermal insulation products. The most popular
thermal insulation products (Figure 9) are:

– expanded (EPS) or extruded (XPS) polystyrene;
– polyurethane foam.

---

[1] More severe requirements are being prepared within a new thermal code expected
for 2006.

Extruded polystyrene

Polyurethane foam

Figure 9.   Examples of thermal insulation products.

In execution

Painted

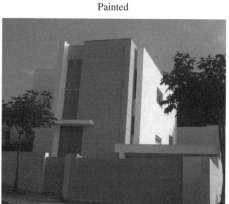

Figure 10.   Examples of traditional render.

3.3   *Damp proof course*

Damp proof course near the soil and flashing are generally made with rich hydrofuge mortar reinforced with fibre glass mesh. The use of metallic, plastic, rubber or asphaltic sheets is not frequent.

3.4   *Wall finishes*

Usually in Portugal the majority of walls were formerly covered with traditional renders made of inorganic binders, aggregates and sometimes admixtures for improving water proofing. These renders were normally painted (Figure 10), but in the last few years alternative solutions have had significant expression.

The problem of maintenance costs, the need to improve wall resistance to rain penetration and the difficulty of hiring skilled workers to make renders, have led to the development of alternative solutions:

– ceramic and thin stone tiles bedded on the wall render with cement based adhesive is perhaps, today, one of the most frequent solutions; these elements, mainly ceramic tiles, are available in many different sizes and aspects; some of them have the aspect of ceramic facing bricks (Figure 11);
– factory-made rendering materials with pigments, coloured sand or small pieces of stone, applied in a single coat (one coat render) (Figure 12);
– cavity walls with facing units (Figure 13);
– external thermal insulation composite systems (ETICS) bonded onto the wall, with a special rendering consisting of one or more layers (site applied), one of which is reinforced with fibre glass mesh (Figure 14).

Other different wall finishes are used but with low expression in residential buildings.

Ceramic tiles   Granite stone tiles

Figure 11. Examples of tiles.

Figure 12. Example of a factory made coat render.

Facing clay units in execution

Facing clay units finished

Split dense aggregate concrete units

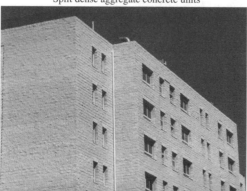

Figure 13.   Examples of cavity facing unit walls.

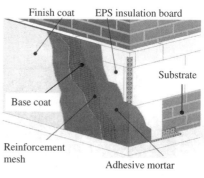

Figure 14.   Example of external thermal insulation composite system (ETICS).

3.5   *Most common masonry building enclosures*

In Portugal, building regulations, concerning energy economy and heat retention [8], protection against noise [12] and safety in case of fire [13] dictate, to some

Typical housing building structures and cavity clay unit walls

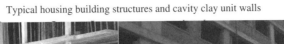

Box for roller blind   Lintel

Single leaf wall in lightweight concrete blocks

Figure 15.   Examples of current Portuguese masonry buildings enclosures.

extent, the solutions used for wall enclosures. At the moment, the most popular
masonry building enclosures (Figure 15), of simple infilling, are:

– Cavity walls made with clay units of high horizontal perforation, of which the
  thickest leaf usually does not exceed 0.15 m width, with thermal insulation in
  the cavity (generally expanded or extruded polystyrene boards). These walls
  are generally poorly cared concerning wall ties, damp proof barrier, insulation

fixing, thermal bridges and structural connections. The finishes are various as referred to on 3.4.

- Cavity walls with outer leaf made of facing clay or concrete units and inner leaf made of clay units of high horizontal perforation, generally with 0.11 or 0.15 m width. The cavity face of the inner leaf has improved waterproofing using a cementitious or synthetic painting, in some cases with a reinforcing mesh. The thermal insulation is made of expanded or extruded polystyrene boards or polyurethane foam. The wall ties and thermal bridge protection, when present, are not well cared. Some weepholes are generally foreseen.
- The use of concrete blocks occurs in a few small constructions sometimes with structural contribution in confined masonry solution. In this case, cavity walls with the inner leaf of clay brick are used.

The use of single leaf wall solution has been growing but is not yet frequent. Lightweight aggregate concrete, autoclaved aerated concrete and lightweight clay blocks are used. These units, with the exception of autoclaved aerated concrete blocks, are vertically perforated and their thickness is usually between 0.25 and 0.30 m.

In these last solutions, if the width of the unit is higher than 0.15 m, normally the units will have interlocking features (tongue and groove system), and consequently vertical joints are not filled with mortar.

3.6   *Existing problems*

a)   Design and construction
The design of non-structural walls is entrusted to architects, but generally there is not an adequate specification of masonry works concerning the type of materials, characteristics, details of execution and singular points. As the buildings structures and installations are correlated with masonry works, there is a special need for compatibilization that normally is not correctly done.

Furthermore the designer has some difficulties in specifying the construction requirements because no code of practice exists for these works. So, it is usual to define on site, during the construction process, the effective quality of masonry walls. With a good design, if the contractor prepares correctly the work and the workers are skilled, the masonry will have quality. However, without harmonized mechanisms, it is very difficult to control and force the contractor to do a good work.

b)   Special difficulties
The main problems detected in Portuguese construction, which are at the origin of an important number of cases of pathology, are:

- Reinforced concrete slabs excessively deformable, producing mechanical actions and cracking of masonry;
- Connections between walls (including renders) and structure not correctly solved; the problems are aggravated with some incorrect practices to minimize thermal bridges;
- Cavity walls mechanically weak and incorrectly constructed considering cavity cleaning, installation of ties, placement of weepholes, position and fixing of thermal insulation;

Figure 16. Examples of construction defects and masonry pathology.

- Singular points around openings not studied and generally solved on site with too much improvisation;
- Finishes, renders and tiles, chosen without a technical evaluation and applied too fast;
- Architectural solutions for the façades not taking in account the incidence of rainfall, the workmanship quality and the need for durability.

Figure 16 shows examples of construction defects and masonry pathology.

4 EVOLUTION TRENDS

4.1 *Solutions for existing problems*

Masonry walls are one of the most important construction elements present in buildings, particularly exterior walls that separate the internal and external environments, and they must contribute adequately to the functions of building envelopes.

Figure 17.   Building pathology according to a statistic study in France [14].

Unfortunately masonry gives an important contribution to building pathology, as is shown in a statistic of another country [14] that is assumedly similar to the Portuguese situation (Figure17).

Usually, the design of non-structural walls and the responsibility for ensuring the environmental adequacy of buildings has been entrusted to architects, including coordination and integration of the several aspects, but today this work requires the cooperation of many experts. This objective is nowadays more relevant, having in mind that users demand the construction "products" to have similar guaranties to the generality of other products available on the market.

In Portugal, this mentality is beginning to change on the part of owners, designers and contractors, the relenting of new building constructions and a significant emphasis on quality aspects should be important contributions to these changes.

In order to achieve external walls able to fulfil the requirements of building envelopes for the climatic conditions of the building place, the mastering of different areas of knowledge and their interactions are very important:

– Building sciences – raintightness, condensation and thermal performances, acoustic performances and fire safety;
– Structural safety of load-bearing or infilling masonry;
– Masonry materials – different units, mortars, reinforcements, ancillary components and combined behaviour;
– Construction technologies – workmanship and construction practices.

The awareness of the need of a complete masonry design is indispensable, including details solving singular points of the walls, considering masonry materials characteristics and construction practices.

4.2   *Trends regarding labour, quality, durability and productivity*

Masonry walls and their various components should be designed considering their overall cost, including construction, operation and maintenance.

The effort to rationalize masonry construction and laying with productivity profit and being less painful to workers is indispensable. The success of these efforts requires a deep knowledge of masonry behaviour. The main evolutions in these domains are:

– development of larger units, with good thermal insulation that could be used in some countries in single leaf walls without complementary thermal insulation, as alternative to cavity walls;

– units with tongue and groove joints to help the laying;
– units with shapes that facilitate filling and handling;
– use of thin mortar joints if the tolerances of the units are compatible.

The workmanship characteristics available for masonry works are changing. The availability of skilled workers subjected to practical long training periods has reduced a lot, being replaced by unskilled workers. Furthermore, the quick pace of construction works existing nowadays, and some current architectural solutions make the buildings more vulnerable to workmanship quality.

These particularities increase the importance of a good detailing of the works and of the adoption of simple solutions, less dependent on workmanship. These concerns are extendable to mortars used in masonry works and to finishes.

The growing concern for environmental aspects is pressing the construction activity towards sustainability represented by the adoption of more natural and less aggressive solutions.

Building materials used in the construction industry should not be harmful for the environment and for the human being. Durable or reusable building materials that minimize the use of natural resources should be preferred in order to minimize pollution. The investment in the acquisition of this type of materials will be largely compensated by its longer life and less waste in the long run.

4.3   *New developments*

The choice for a certain masonry solution and unit depends not only on its functional performance, but also on the analysis of other aspects like equipment, stock facilities, cost of work and necessary workmanship qualification.

The arrival of new building techniques and technology implies the acquisition of new work concepts. In Portugal, the transitory character of employment in construction and the lack of basic and continuous training are a barrier to the desired construction quality. In this sense, it is advantageous, at the moment, to use simpler building techniques.

The use of better quality clay and concrete units with accurate dimensions and special shape units for masonry singular points results in a better and less heavy work, offering better conditions to workmanship and better performances to building companies.

The research about the geometry and the material used in the production of masonry units and the optimising of mechanical and thermal properties are important. The mixing of granular materials in clay that vaporises during furnace cooking and the use of lightweight concrete in masonry units makes it possible to obtain lighter and more insulating masonry units with acceptable mechanical resistance. The use of those elements in Portugal has some expression and investment from the industry is being done in this field.

Single-leaf masonry external building walls, that in part dispense skilled workmanship inherent to cavity walls, can be one of these techniques with minor probability of occurrence of construction pathologies. From a point of view of economy, an analysis of current solutions of external building walls with the same coefficient of heat transmission, shows that the solutions of single leaf walls can be cheaper than the solutions of cavity walls [15].

5   SIGNIFICANT REFERENCES AND DOCUMENTATION

[1] Afonso et al. – *O Sector da Construção. Diagnóstico e eixos de intervenção.* Lisboa, Observatório das PME's, IAPMEI, 1998 (in Portuguese).

[2] INE – Instituto Nacional de Estatística – *Estatísticas da Construção, 1997.* Lisboa, INE (in Portuguese).

[3] Sousa, H. et al. – *Rain watertightness of single leaf and cavity walls.* In Proceedings of International Symposium on Moisture Problems in Building Walls. Edited by Vasco Freitas e Vítor Abrantes. Porto, FEUP, 1995 (in Portuguese).

[4] Santos, Fernando – *Alvenarias em Edifícios. Inventariação das Soluções Utilizadas e Proposta de um Novo Sistema.* Tese de Mestrado. Porto, FEUP, 1998 (in Portuguese).

[5] STATEC – Service Central de la Statistique et des Études Economiques – *Les bâtiments achevés en 2002.* Bulletin du Statec 4/2004. Luxembourg, STATEC, 2004 (in French).

[6] ISTAT – Istituto Nazionale di Statistica – *Statistiche dell'Activitá Edilizia. Anno 2001 – Dati provisori.* Informaizioni n° 32 – 2003. Roma, ISTAT, 2003 (in Italian).

[7] SBi – Danish Building Research Institute – *Statistic data on structural systems for buildings.* Information received by letter. Horsholm, SBi, 2002.

[8] *Regulamento das Características de Comportamento Térmico dos Edifícios.* Decreto – Lei n°. 40/90, de 6 de Fevereiro (in Portuguese).

[9] LNEC – *Inquérito à Produção Nacional de Materiais para Alvenaria.* Lisboa, LNEC, 1986 (in Portuguese).

[10] Rei, João – *Edifícios de Pequeno Porte em Alvenaria Resistente. Viabilidade Técnico-económica.* Tese de Mestrado. Porto, FEUP, 1999 (in Portuguese).

[11] Serra e Sousa, A.; Silva, R. et al. – *Manual de Alvenaria de Tijolo.* Coimbra, APCER, CTCV, FCTUC, 2000 (in Portuguese).

[12] *Regulamento dos Requisitos Acústicos dos Edifícios.* Decreto – Lei n° 129/2002, de 11 de Maio (in Portuguese).

[13] *Regulamento de Segurança Contra Incêndio em Edifícios de Habitação.* Decreto–Lei n°. 64/90 de 21 de Fevereiro (in Portuguese).

[14] Bureau Securitas – *Étude statistique de 12200 cas de sinistres survenus en 1982.* Annales de l'ITBTP, N° 426, Séries Questions Generales 162, Paris, Juillet-Aôut 1984 (in French).

[15] Alves, Sérgio – *Paredes Exteriores de Edifícios em Pano Simples. Fundamentos, Desempenho e Metodologia de Análise.* Tese de Mestrado. Porto, FEUP, 2001 (in Portuguese).

CHAPTER 12

Veneer walls in seismic areas in U. S. A.[1]

J. Gregg BORCHELT

Vice President
Brick Industry Association
Reston
Virginia
U.S.A.

ABSTRACT

This paper describes anchored brick veneer as used in the United States of America and provides an explanation of the model building codes in use there. The prescriptive requirements as found in the model building codes used in the USA are included. The seismic design provisions for anchored brick veneer in model building codes, including those in the residential code, are explained. Performance of veneers and related structural elements under seismic loading is explained.

1 INTRODUCTION

Brick veneer is a popular form of exterior wall cladding in all parts of the United States of America. Though used primarily in residential construction, it is also used extensively in retail, manufacturing, institutional and office buildings. Applications range from single story homes to high-rise structures. The majority of brick is used in the eastern United States, a location that has not historically incorporated seismic activity in design of buildings.

Building code requirements for brick veneer evolved empirically, with minor changes coming about as designers examined the performance of brick veneer. These changes include brick veneer requirements for performance related to seismic forces.

The Federal Government's mandate to reduce property loss and deaths due to seismic events had a significant effect on the use of brick veneer in the residential market.

---

[1] This paper have been presented at the Conference SISMICA2004, organized by the University of Minho (Portugal), in April 2004. It is reproduced in this publication, with the permission of the author and of the organizers, slightly modified.

This mandate increased seismic loads and changed the design and construction requirements for all materials, including limits on the height of brick veneer over wood frame backing. The brick and masonry industries have responded to these restrictions with analytical studies and resulting changes to the requirements for brick veneer and wood frame.

This paper covers only anchored brick veneer, the more widely used type of masonry veneer. Similar elements of veneer construction as found in the various model building codes are covered in the following subsections.

2   BACKGROUND ON BRICK VENEER

2.1   *Definitions*

Veneer: *A facing attached to a wall for the purpose of providing ornamentation, protection, or insulation, but not counted as adding strength to the wall.*
Backing: *The wall or surface to which the veneer is secured.*

These definitions of **veneer** and **backing** taken from the 2000 International Building Code, or slight modifications, are found in the model building codes used in the United States. These definitions present several conditions that imply how all veneer materials, including brick, are treated with respect to design and performance.

First, the veneer is not intended to resist loads; it is nonstructural. Loads imposed on the veneer are simply transferred to other structural materials designed to resist these loads. This backing resists these loads and transfers them to the structural frame. The exterior surface may be any of a variety of materials, each with specific requirements for thickness and attachment. The prescriptive requirements for attachment are often based on intuition, not engineering. Typically there is a provision in the code that the entire wall assembly must provide a weather-resistant barrier.

When the facing is masonry, it is known as masonry veneer. When the facing is brick it is known as brick veneer. Reasons for using brick include the variety of colors, sizes and textures available and the durability and low maintenance of brickwork. Brick veneer is of one of two types: adhered or anchored.

In adhered veneer the brick is attached to the backing by the adhesive force of glue or a cement-based material to a continuous surface on the backing. Adhered brick veneer is not frequently used. Anchored brick veneer (Figure 1) is attached to the backing by a series of regularly spaced steel ties or anchors. There is an air space between the inside surface of the veneer and the outside surface of the backing. The airspace, along with flashing and weep holes, reduce water penetration into the building.

2.2   *Brick and mortar*

Brick must meet the requirements of American Society for Testing and Materials Standards C 216 or C 652. These standards are for facing and hollow brick, respectively. They contain requirements for freeze-thaw durability and appearance. Brick units used in anchored veneer are required by model building codes to be at least 54 mm (2 5/8 in.) or 50 mm (2 in.) thick. Actual brick thicknesses of 76 mm (3 in.), 90 mm (3-½ in.), and 95 mm (3 5/8 in.) are typically used.

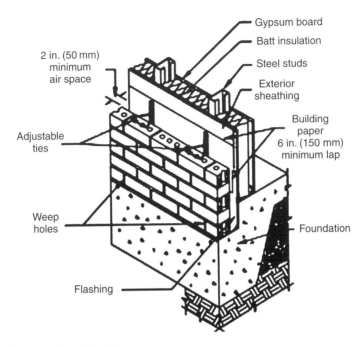

Figure 1.   Anchored brick veneer.

Mortar must conform to the requirements of ASTM C 270. This standard includes a wide variety of permitted cementitious materials and sand. Mortar is specified by proportions of materials or by properties of laboratory tested mortar. Lower strength cementitious materials and mortar types are prohibited from use in the higher seismic areas by model codes. Mortar Type N is typically used with brick veneer.

2.3   Anchors

The anchors specified in each of the model codes are similar. They are of steel and may be:

– corrugated sheet of 0.76 mm (No. 22 gauge) by 22 mm (7/8 in.),
– sheet metal of 1.5 mm (0.06 in.) by 22 mm (7/8 in.), or
– wire of minimum MW11 (No. 9 gauge).

Anchors come in a variety of configurations (Figures 2 and 3). Corrosion resistance is achieved with zinc or epoxy coatings or the use of stainless steel. Play in two piece adjustable anchors is limited to 1.6 mm (1/16 in).

2.4   Backing

The structural backing used with brick veneer may be a continuous surface such as concrete masonry or concrete, or it may be steel or wood framing (studs) with an exterior sheathing. Wood studs are most often used in residential construction. Most backings with sheathing also require a weather-resistive barrier on its exterior.

Figure 2.   Typical unit veneer anchors.

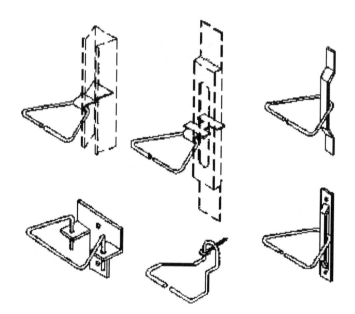

Figure 3.   Typical adjustable anchors.

2.5   *Construction of brick veneer*

The backing is in place before bricklaying begins. This is true even if the backing is masonry. In the United States a separate construction trade will complete non-masonry backings. This includes attaching all or at least one part of the anchors. Components of the wall system, such as door and window frames, a weather-resistive barrier and flashing, are installed prior to brick being laid.

Brickwork is laid to a line, with full bed and head joints required with solid brick and face shell bedding required with hollow brick (Figure 4a and b). The air space between the brickwork and the backing is to be kept clear of mortar droppings and debris.

Proper installation of anchors is, of course, necessary. The anchor must be embedded into the mortar joint in order to develop its strength. Wire anchors must have at least 25 mm (1 in.) of cover to each face of the brickwork. Plate anchors must penetrate at least one-half of the veneer thickness.

Size and type of connection of the anchor to the backing must meet the requirements of the building code and be installed in accordance with the anchor

(a)   (b)

Figure 4.   Construction of brick veneer.

manufacturer's recommendations. Obviously nails and screws must be attached to the studs, not the sheathing.

3   BUILDING CODE REQUIREMENTS FOR ANCHORED VENEER

3.1   *Building codes in the United States*

There is no government mandated building code in the United States of America. Rather, building codes are adopted into law at the local level by a state, a county, or a city. These governmental jurisdictions typically adopt a model code with local amendments.

Model codes are prepared by a modified consensus process, with code officials determining the content and changes.

Model building codes contain the following primary parts:

– Administrative issues that relate to adoption, building use and occupancy;
– Fire-resistive construction and fire protection;
– Accessibility and environmental control;
– Structural design loads, tests, and inspection;
– Soils and foundations;
– Material specific requirements for structural design and construction;
– Electrical, mechanical, plumbing, and conveying systems.

The original content of these primary parts comes from a variety of sources. The first three of these parts are typically prepared by model building code writing groups. Organizations representing design professionals, typically the American Society of Civil Engineers, write the standards that relate to structural design loads, tests, and inspection and the part on soils and foundations.

As would be expected, organizations representing the various material groups write the standards that include the requirements for those materials. Organizations representing design professionals also write the standards for electrical, mechanical, plumbing, and conveying systems. These standards are written under a consensus system and are adopted by reference with modifications by the model building codes.

Changes to these codes are accepted on a cyclic basis. Changes come about as a result of changes in the referenced standards and changes introduced by an individual or organization.

Recently the federal government has been active in supporting changes to model codes, specifically with respect to seismic performance. These changes are presented at the code hearings where the proponent and opponents express their views. Final action is taken by the code officials who are members of the model code organization. At this time, a new edition of most model codes is approved on a three year cycle. Amendments are accepted once during that three year cycle.

3.2   *Model building codes*

There has been a significant change in model building codes in effect in the United States in recent years. Three model code organizations, that each promulgated a separate model building code used with regional preference, have merged into the International Code Council.

Thus three model building codes and a common residential code have been replaced with a single code for one- and two-family dwellings, the International Residential Code [1], and one for other structures, the International Building Code [2]. A second model building code, the National Fire Prevention Association 5000 has just been approved. This group also uses the International Residential Code.

Both of these model building codes refer to the *Building Code Requirements for Masonry Structures* [3]. This document is written by the Masonry Standards Joint Committee and covers all masonry construction. It is a "joint" effort of the American Concrete Institute, the American Society of Civil Engineers, and The Masonry Society. This will be referred to as the MSJC Code in this paper. The MSJC Code includes requirements for structural masonry and veneer.

The veneer sections of model building codes provide for two means of compliance: a design using engineering principles or a set of prescriptive materials and practices that have proven to result in veneer that performs properly. Almost all veneer is designed using the prescriptive requirements. The following sections provide background on the current prescriptive requirements for anchored veneer, with emphasis on those related to seismic performance.

3.3   *Federal mandate to reduce earthquake damage*

The United States government established a mandate to reduce the effects of earthquakes on buildings. The National Earthquake Hazard Reduction Program (NEHRP) began in 1977 and addresses both the design loads associated with earthquakes and the resistance to them provided by building materials and elements. This program has developed recommended provisions [4] for seismic design which are taken to the model building codes for adoption.

These provisions cover all material types. Provisions for anchored veneer refer to the MSJC Code. Thus there is nearly consistency in application of seismic requirements for anchored veneer in the model building codes. There is an inconsistency because of the differences between the International Residential Code and the International Building Code. These will be explained in more detail as the content of the codes is discussed.

3.4 *First seismic requirements for veneer*

Prescriptive requirements for anchored veneer first included seismic provisions in the 1982 edition of the Uniform Building Code [5]. These requirements apply to anchored veneer made with masonry units of 127 mm (5 in.) maximum thickness that is anchored directly to structural masonry, concrete, or studs. Seismic requirements are underlined:

1. Anchored veneer shall be supported on footings, foundations or other non-combustible support.
2. Where anchored veneer is applied more than 7.6 m (25 ft) above adjacent ground elevation, it shall be supported by noncombustible, corrosion-resistant, structural framing having horizontal supports spaced not over 3.7 m (12 ft) vertically above the 7.6 m (25 ft) height.
3. Noncombustible, noncorrosive lintels and noncombustible supports shall be provided over all openings where the veneer unit is not self-spanning. The deflections of all structural lintels and horizontal supports shall not exceed 1/500 of the span under full load of the veneer.
4. Anchors shall be corrosion resistant, and if made of sheet metal, shall have a minimum size of 0.76 mm (No. 22 gauge) by 25.4 mm (1 in.) or, if made of wire, shall be a minimum of MW11 (No. 9 gauge).
5. Anchor ties must be spaced to support not more than 0.19 m^2 (2 ft^2) of wall area but not more than 0.61 m (24 in.) on center horizontally.
6. *In Seismic Zones No. 3 and No. 4 anchor ties shall be provided to horizontal joint reinforcement wire of MW11 (No. 9 gauge) or equivalent.*
7. *The joint reinforcement shall be continuous with butt splices between ties permitted.*
8. When used over stud construction, the studs shall be spaced at a maximum of 406 mm (16 in.) on centers.

Seismic Zones 3 and 4 were the highest of four zones in effect at that time.

A subsequent modification to these requirements, made in 1991, required that the joint reinforcement be mechanically attached to the veneer anchors and that butt splices in the reinforcement were permitted between the anchors.

3.5 *Masonry Standards Joint Committee Code*

a) General
Requirements for anchored veneer were introduced into the *Building Code Requirements for Masonry Structures* the 1995 edition. These came from a compilation of veneer requirements in the previous three model building codes. Thus, the UBC

requirements were used as the basis for seismic provisions. The MSJC Code is included by reference in the International Building Code.

Key prescriptive provisions from MSJC 1995 Code and subsequent editions through 2002 for anchored veneer are:

"**6.2.2.1** Prescriptive requirements for anchored masonry veneer shall not be used in areas where the basic wind speed exceeds 110 mph (177 km/hr) as given in ASCE 7.

6.2.2.3 Vertical support of anchored masonry veneer

6.2.2.3.1 The weight of anchored veneer shall be supported vertically on concrete or masonry foundations or other noncombustible structural supports, except as permitted in Sections 6.2.2.3.1.1, 6.2.2.3.1.4, and 6.2.2.3.1.5.

6.2.2.3.1.1 Anchored veneer is permitted to be supported vertically by preservative-treated wood foundations. The height of veneer supported by wood foundations shall not exceed 18 ft (5.49 m) above the support.

6.2.2.3.1.2 Anchored veneer with a backing of wood framing shall not exceed the height above the noncombustible foundation given in Table 6.2.2.3.1.

6.2.2.3.1.3 If anchored veneer with a backing of cold-formed steel framing exceeds the height above the noncombustible foundation given in Table 6.2.2.3.1, the weight of the veneer shall be supported by noncombustible construction for each story above the height limit given in Table 6.2.2.3.1.

Table 6.2.2.3.1. Height limit from foundation.

Height at plate, ft (m)	Height at gable, ft (m)
30 (9.14)	38 (11.58)

6.2.2.3.1.4 When anchored veneer is used as an interior finish on wood framing, it shall have a weight of 40 lb/ft² (1915 Pa) or less and be installed in conformance with the provisions of this Chapter.

6.2.2.3.1.5 Exterior masonry veneer having an installed weight of 40 psf (195 kg/m²) or less and height of no more than 12 ft (3.7 m) is permitted to be supported on wood construction. A vertical movement joint in the masonry veneer shall be used to isolate the veneer supported by wood construction from that supported by the foundation. Masonry shall be designed and constructed so that masonry is not in direct contact with wood. The horizontally spanning element supporting the masonry veneer shall be designed so that deflection due to dead plus live loads does not exceed l/600 nor 0.3 in. (7.6 mm).

6.2.2.3.2 When anchored veneer is supported by floor construction, the floor shall be designed to limit deflection as required in Section 1.10.1.

6.2.2.3.3 Provide noncombustible lintels or supports attached to noncombustible framing over all openings where the anchored veneer is not self-supporting. The deflection of such lintels or supports shall conform to the requirements of Section 1.10.1.

6.2.2.9 Veneer laid in other than running bond. Anchored veneer laid in other than running bond shall have joint reinforcement of at least one wire, of size W1.7 (MW11), spaced at a maximum of 18 in. (457 mm) on center vertically."

Table 1.   Veneer Anchor Frequency Requirements, MSJC Code.

Anchor Type	Area of Veneer per Anchor	
	Square Meters	Square Feet
Corrugated Sheet, Single Piece	0.25	2.67
Flat Plate, Single Piece	0.33	3.5
Wire, MW 11 (W1.7) and smaller, Single Piece	0.25	2.67
Wire, larger than MW 11 (W1.7), Single Piece	0.33	3.5
Two Piece Adjustable	0.25	2.67

There are specific requirements for anchors to be used with anchored brick veneer. Veneer anchors are identified by steel item used to fabricate them. These include corrugated sheet metal, sheet metal, wire, and joint reinforcement.

Anchors are further identified as single or adjustable (two-piece) anchors. Adjustable anchors must not disengage and there is a limit of clearance between parts of 1.6 mm (1/16 in.). Pintle anchors must have two legs of wire size MW1.8 (W2.8) and an offset not exceeding 31.8 mm (1¼ in.).

These classifications determine the frequency requirements. Table 1 summarizes the area of veneer required for typical ties. Anchors are spaced at a maximum of 813 mm (32 in.) horizontally and 457 mm (18 in.) vertically. Additional anchors are placed around openings larger that 406 mm (16 in.) in either direction. The MSJC Code includes specific requirements for placing the anchor in the mortar joints.

The type of backing used with the brick veneer determines a number of related wall components. Anchor type is one of these items. Corrugated anchors are permitted only with wood backing. Thus, wood stud backing permits any anchor type. Steel stud and concrete backing require adjustable anchors.

Masonry backing can accept wire or adjustable anchors or joint reinforcement. The minimum and maximum air space dimensions between the inside face of the brick and the outside face of the backing or sheathing are set.

b)   Seismic requirements

Specific criteria for seismic conditions are introduced and are cumulative with increasing seismic activity. The four Seismic Zones used in the previous codes were updated to five Seismic Design Categories (SDC).

The building's Seismic Design Category is established by the use of the building and severity of the design earthquake ground motion at the site. Alphabetical designations for SDC begin at A and end with F as the highest.

Seismic Design Categories A and B
There are no specific seismic requirements.

Seismic Design Category C

"**6.2.2.10.1.2** Isolate the sides and top of anchored veneer from the structural so that vertical and lateral seismic forces resisted by the structure are not imparted to the veneer."

Figure 5.   Anchored veneer with ties and joint reinforcement.

Seismic Design Category D

"**6.2.2.10.2.2** Support the weight of anchored veneer for each story independent of other stories.
6.2.2.10.2.3 Reduce the maximum wall area supported by each anchor to 75% of that required for the specific tie used. Maximum horizontal and vertical spacing are unchanged.
6.2.2.10.2.4 Provide continuous, single-wire joint reinforcement of minimum wire size MW11 (W1.7) at a maximum spacing of 457 mm (18 in.) on center vertically."

Seismic Design Categories E and F

"**6.2.2.10.3.2** Provide vertical expansion joints at all returns and corners.
6.2.2.10.3.3 Mechanically attach anchors to the joint reinforcement with clips or hooks. (Figure 5)"

As indicated in the Commentary to the MSJC Code [3], these requirements are to provide added flexibility to the veneer. The Commentary states:

"**6.2.2.10** Requirements in seismic areas – These requirements provide several cumulative effects to improve veneer performance under seismic load. Many of them are based on similar requirements found in Chapter 30 of the Uniform Building Code6.14. The isolation from the structure reduces accidental loading and permits larger building deflections to occur without veneer damage. Support at each floor articulates the veneer and reduces the size of potentially damaged areas. An increased number of anchors increases veneer stability and reduces the possibility of falling debris. Joint reinforcement provides ductility and post-cracking strength. Added expansion joints further articulate the veneer, permit greater building deflection without veneer damage and limit stress development in the veneer."

Modifications to these seismic requirements were introduced in the 2005 edition. The following changes have been made:

1. The requirement to support the weight of the veneer for each story independent of other stories was removed from Seismic Design Category D and implemented in SDC E and F.
2. The requirement for continuous, single wire joint reinforcement was removed from Seismic Design Category D and implemented in SDC E and F.

Both of these changes were based on engineering investigations. Further explanation is given in the section: "VENEER PERFORMANCE UNDER SEISMIC LOADS".

3.6   *International Residential Code*

This separate building code [1] for one-and two-family dwellings is more pre-scriptive to its overall philosophy to design and construction. It covers limited types of buildings, as the name implies.

The smaller scope of this charge, and the desire to limit the amount of engineering input, results in the inclusion of derived design tables for wood and steel stud construction. One significant change is to divide SDC D into two subzones, D_1 and D_2, separated at a Short Period Design Spectral Response Acceleration of 0.83 g.

The IRC also includes specific prescriptive requirements for anchored veneer. It does not reference the MSJC for veneer requirements. The prescriptive requirements for anchored veneer, with relation to the MSJC code requirements, are:

1. Requirements for a weather-resistant envelope, including a weather-resistive barrier, drainage means, and flashing.
2. Minimum nominal thickness of brick veneer of 50 mm (2 in.).
3. Anchor materials and types are MW11 wire and corrugated sheet metal of the same size as in the MSJC Code.
4. Veneer anchors frequency is one anchor per 0.0.302 m^2 (3.25 ft^2) for all anchor types.
5. Maximum spacing for anchors is 601 mm (25 in.) horizontally.
6. Backing types are the same and have the same permitted anchors and air space dimensions as in the MSJC Code.
7. Anchored brick veneer with wood or cold-formed backing is limited to a maximum thickness of 127 mm (5 in.).
8. Support of veneer on wood of cold-formed frame is permitted, with detailing requirements for support on wood.
9. Sizes and spans for steel and reinforced masonry lintels are given.

Cumulative seismic requirements are:

Seismic Design Categories A and B
Exterior masonry veneer with a backing of wood or cold-formed steel framing shall not exceed 9.14 m (30 ft) in height above the noncombustible foundation, with an additional 2.35 m (8 ft.) permitted for ends.

Seismic Design Category C
In other than the topmost story, the length of bracing shall be 1.5 times the length otherwise required.

Seismic Design Category D_1 and D_2

1. Exterior masonry veneer with a backing of wood or cold-formed steel framing shall be limited to the first story above grade.
2. Each tie shall support not more than $0.186\,m^2$ ($2\,ft^2$) of wall area.
3. Veneer ties shall be mechanically attached to horizontal joint reinforcement wire a minimum of MW11 (9 gage). The horizontal joint reinforcement shall be continuous in the veneer bed joint, with lab splices permitted between the veneer tie spacing.

The seismic performance of wood stud construction is not well understood and many combinations of wall sheathing and nailing schedules are possible. Since veneer is not intended to resist loads, the seismic load from the weight of the veneer is applied to the wood stud shear walls.

Conservative assumptions with respect to the strength of the shear walls indicated that typical wood frame construction in SDC C could not resist the resulting seismic force from brick veneer more than 2.35 m (30 ft) in height. In SDC D, the limit was one story. This change was coupled with an increase in the seismic activity in the central and eastern United States. The result was that a significant portion of the market for anchored brick veneer was threatened.

In an effort to reduce the effect of these seismic restrictions, the brick industry worked with the writers of wood industry provisions. The first result was to provide a strengthened wood shear wall system that was capable of resisting the seismic load from the veneer. Tie downs were required at the end of shear walls; a prescribed sheathing material and nailing schedule were added as seismic loads increased.

These requirements were added to the International Residential Code in 2002 amendments to the 2000 edition. Further work for the 2003 edition of the NEHRP Provisions took a more realistic look at the performance of brick veneer. The fact that the veneer is capable of resisting loads was recognized and the limits on the permitted height of anchored masonry veneer were reduced.

This activity and other amendments have been adopted in the 2003 IRC as follows:

- Anchored veneer tie frequency was reduced one anchor per $0.0.248\,m^2$ (2.67 ft^2).
- A maximum slope of roofs supporting masonry veneer was established at 7:12 without special construction. Roofs sloped up to 12:12 required steel stops on the supporting angles.

Seismic Design Category D_1 and D_2
The requirement for joint reinforcement was eliminated. The permitted height of anchored veneer was increased to 6.10 m or 9.14 m (20 ft or 30 ft), plus 2.44 m (8 ft) in the gabled ends. These greater heights depend on veneer leaf thickness, concrete or masonry backing for the lower 3.05 m (10 ft), and the presence of specific braced panels and hold down connectors.

4   VENEER PERFORMANCE UNDER SEISMIC LOADS

4.1   *Perception*

Media coverage of seismic events invariably includes images of fallen brick and the damage they cause. Photos of X-type cracking in masonry elements between windows and piles of masonry rubble on the sidewalks are often shown. There is seldom any explanation as to the type of element that failed or the age or the prior condition of the building.

The impression is that masonry construction is not appropriate for use in seismically active areas. It is true that some existing brick masonry buildings, both bearing wall and veneer, are not adequate to resist potential seismic forces.

However, there are design and construction procedures that will upgrade existing structures to satisfactory performance. Further, post-earthquake investigations have shown that current design and construction procedures for brick structures, including veneers, provide resistance to seismic forces [6, 7, 8, 9].

4.2   *Load generation and resistance*

Loads generated by earthquakes are caused by the inertia of the object subjected to ground shaking. The seismic load may occur in any direction, resulting from the direction of movement of the ground.

For analytical purposes, seismic loads are typically assumed to be applied in the plane of the wall and perpendicular to the plane of the wall. The load in any element is directly proportional to weight of that element, and the load occurs with the element itself. This means that the basic assumption that veneer is not intended to resist loads is violated.

Brick veneer is subject to seismic loads and it must be able to resist those loads and transfer them to the structural system of the building. Analytical studies have shown that anchored veneer does resist seismic loads and it contributes to the resistance of seismic loads [10].

Seismic loads due to the weight of the wall from in-plane movement are resisted by the brick veneer only. There must be sufficient masonry present to develop the necessary shear resistance without cracking, or it must be reinforced. Transfer of in-plane forces to the backing will depend on the type of veneer anchor used.

Loads perpendicular to the plane of the wall are resisted by both the veneer and the backing. These develop flexural tension in the masonry and it spans as a beam supported on elastic supports (the anchors). The anchors transfer those reactions perpendicular to the face of the masonry from the masonry to the backing. The strength of the brick veneer, the anchors, and the backing must be sufficient to withstand the loads.

4.3   *Failure mechanisms*

a)   In-plane loads
In-plane loads in the veneer generate corresponding shear stress. Sliding may take place at the support since the presence of flashing provides a weakened plane. In-plane stresses may cause a series of stepped or horizontal cracks in the veneer if

Figure 6.   In-plane failure.

the applied force is greater than the shear resistance of the masonry. X-type crack-ing is prevalent near and between openings due to load reversal.

Tall, thin masonry elements may develop horizontal cracks in bed joints near the top and bottom of the elements. If the masonry elements are not strong enough, and if the earthquake is severe, collapse of the veneer can occur [6, 7, 8, 9] (Figure 6).

Johnson and McGinley [11] have shown that there is typically sufficient masonry to develop the necessary shear resistance without cracking. Further, the corrugated steel anchors typically used in residential construction provide signif-icant transfer of load, if needed, and do so with ductile behavior.

b)   Out-of plane loads

Since the brickwork, anchors and backing work together to resist out-of-plane loads, the resistance mechanism is quite complex. The brickwork has a high flex-ural rigidity compared to stud backing and initially resists the majority of the load. A crack will develop in a bed joint of the masonry if the wall construction is not sufficient to resist the load.

The flexural tensile stress in the brick veneer reaches its maximum value just prior to cracking. At this state anchors nearest to rigid supports may reach their highest load. Immediately after the veneer cracks the portions of the backing away from the rigid supports receive a much higher load. Further, the loads in the anchors are reduced and are distributed more evenly [12].

After this initial crack the maximum flexural stress in the veneer perpendicular to the bed joints is now greatly reduced and may not reach cracking level again. Once a horizontal crack forms the veneer may now span horizontally, with the

Figure 7.   Out-of-plane failure.

vertical studs acting as supports. McGinley, *et al*, [13] have shown that the veneer has sufficient strength to resists these resulting stresses parallel to the bed joints.

Joint reinforcement is detrimental to out-of-plane loading. It reduces the tensile stress in the bed joints and may lead to early cracking and instability [14].

Increasing seismic loads can result in anchor failure, backing failure, or collapse of the veneer (Figure 7). Unbraced vertical sections of masonry can develop high flexural stresses in the veneer and high forces in the top level of anchors. Parapets and gables of veneer construction are susceptible to collapse if support is not provided at the top.

c)   Combined loads

The transition between in-plane and out-of-plane loading, which occurs at corners and returns, causes a unique situation. The veneer at these locations, if bonded around the return, is subject to loads of different magnitudes and types.

Further, the resistance provided by the masonry may be substantially different on each elevation. Determination of the actual stress near returns is difficult. Cracking frequently occurs at the corners. Thus the need for expansion joints near returns [8].

4.4   *Factors influencing failure*

As implied in "Failure Mechanisms", there are several contributing factors to any failure of brick veneer under seismic loading. The height, length, and thickness of the veneer element are certainly critical. They determine the load in the veneer leaf, the amount of material available to resist the load, and the level and type of stress generated.

Thus the location of vertical expansion joints and of openings are of concern with respect to in-plane loads. Location of horizontal expansion joints must also be considered. If these are present the veneer can accept more building drift.

The type of brick and mortar used in the construction will also influence performance, but to a lesser degree. Resistance to in-plane load is determined by material properties.

Higher brick compressive strength and higher mortar cement content increase masonry shear resistance. However, the increase is not directly proportional. The resistance to out-of-plane load perpendicular to the bed joints is relatively low regardless of material combinations when compared to stress generated.

The type of anchor selected has a great influence on the ability of the wall system to resist out-of-plane load. Stronger, more rigid anchors will reduce the load resisted by the brickwork. Weaker, more flexible anchors place greater demand on the brickwork in resisting out-of-plane loads. Resistance to pull-out from or push-through the mortar joint must be high. Anchors are subject to poor performance due to improper installation (Figure 8) and to loss of strength due to corrosion [7, 8, 9].

The backing is important in its contribution to out-of-plane loading. Stiff backings, such as concrete or concrete masonry, share a greater part of load with the veneer than do more flexible backings such as wood or steel studs. Cracking in the veneer is reduced with a stiffer backing. A backing and structural system which permits larger deflection and has a high story drift will also result in larger deflection of the veneer. This will increase cracking, reduce stability of the veneer, and increase anchor loads.

The manner in which the weight of the veneer is supported is also important. Veneer supported at or near each floor level can undergo larger out-of-plane seismic movements. Collapse, if it occurs, is restricted to smaller areas when there is more frequent support of veneer weight. Support at locations above the foundation must be combined with a horizontal expansion joint under these intermediate supports.

Figure 8.   Failure due to pullout of nail from stud.

This joint provides space for movement from frame deflection and story drift to occur. This support at each floor is critical in frame structures and is easily achieved. Support above the foundation is virtually impossible with wood and steel stud structural systems. Bennett [15] reports no technical justification for the requirement to support anchored veneer at each floor level when these light weight backings also act as the structural system.

5 CONCLUSIONS

Anchored masonry veneer continues to be a popular exterior cladding in the United States of America. The building code requirements for its use have progressed from those based on intuition and comparison of masonry cavity walls to those based on engineering analysis.

Seismic performance criteria have evolved in a similar manner. Virtually all post-earthquake investigations show that the performance of brick veneer is determined more by the connection of the anchors to the backing and adherence to proper construction requirements than to the stresses developed in the anchored brick veneer.

The need for inspection of construction, as required in critical facilities, is evident. Adherence to the code required design and construction requirements will maintain confidence in the continued use of anchored brick veneer.

6 REFERENCES

[1] "International Residential Code for One- and Two-Family Dwellings". International Code Council, Country Club Hills, IL, 2000 and 2003.

[2] "International Building Code". International Code Council, Country Club Hills, IL, 2000 and 2003.

[3] "Building Code Requirements for Masonry Structures". ACI 530/ASCE 7/TMS 402 and Commentary on Building Code Requirements for Masonry Structures, The Masonry Society, Boulder, CO, 1995 and 2002.

[4] "NEHRP Recommended Provisions for Seismic Regulations for New Buildings and Other Structures, Part 1 Provisions". Building Seismic Safety Council, Washington DC, 2001.

[5] "Uniform Building Code". International Conference of Building Officials, Whittier, CA, 1982.

[6] Hart, G. C.; Kariotis J.; Noland, J. L. – "Masonry Building Performance. Survey: The Whittier Narrow Earthquake of October 1, 1987". Earthquake Spectra, Volume 4, Number 1, February 1988.

[7] Beiner, R. J.; Vonier T. – "The Loma Prieta, California Earthquake, October 17, 1989: Observations Regarding the Performance of Masonry Buildings". International Masonry Institute, Washington, DC, February, 1990.

[8] Klingner, R. E. – "Performance of Masonry Structures in the Northridge, California Earthquake of January 17, 1994". The Masonry Society, Boulder, CO, June 1994.

[9] Holmes, W. T.; Somers P. – "Northridge Earthquake of January 17, 1994 Reconnaissance Report – Volume 2". Earthquake Spectra, Supplement to Volume 11, Earthquake Engineering Research Institute, Oakland, CA, 1996.

[10] Jalil, I.; Kelm, W.; Klingner, R. E. – "Performance of Masonry and Masonry Veneer Buildings in the 1989 Loma Prieta Earthquake". Proceedings of the Sixth North American Masonry Conference, The Masonry Society, Boulder, CO, June, 1993, pp. 68–692.

[11] Johnson, E. N.; McGinley, W. M. – "The In Plane Shear Performance of Brick Veneer and Wood Stud Walls". Proceedings of the Ninth North American Masonry Conference, The Masonry Society, Boulder, CO, June 2003, pp. 226–237.

[12] Kelly, Goodson T. M.; Mayes, R.; Asher J. – "Analysis of the Behavior of Anchored Brick Veneer on Metal Stud Systems Subjected to Wind and Earthquake Forces". Proceedings of the Fifth North American Masonry Conference, The Masonry Society, Boulder, CO, June 1990, pp. 1359–1370.

[13] McGinley, W. M.; Bennett, R. M.; Johnson, E. M. – "Effects of Horizontal Joint Reinforcement on the Seismic Behavior of Masonry Veneers". Proceedings of the Sixth International Masonry Conference, The British Masonry Society, London, England, November 2002.

[14] McEwen, W. C. – "Private conversation on tests conducted by Masonry Institute of British Columbia", June 2003.

[15] Bennett, R. M. – "Height Limitations for Masonry Veneer in Seismic Category D". Private report for Masonry Alliance for Codes and Standards, February, 2003.